On the (Divine) Origin of Our Species

On the (Divine) Origin of Our Species

DARREL R. FALK

FOREWORD BY Philip Batterham

AFTERWORD BY
Cara Wall-Scheffler & Robert W. Wall

CASCADE *Books* • Eugene, Oregon

ON THE (DIVINE) ORIGIN OF OUR SPECIES

Cascade Books
An Imprint of Wipf and Stock Publishers
199 W. 8th Ave., Suite 3
Eugene, OR 97401

www.wipfandstock.com

PAPERBACK ISBN: 978-1-6667-5701-9
HARDCOVER ISBN: 978-1-6667-5702-6
EBOOK ISBN: 978-1-6667-5703-3

Cataloguing-in-Publication data:

Names: Falk, Darrel R., author.

Title: On the (divine) origin of our species / Darrel R. Falk; Philip Batterham, foreword; Cara Wall-Scheffler and Robert W. Wall, afterword.

Description: Eugene, OR: Cascade Books, 2023 | Includes bibliographical references.

Identifiers: ISBN 978-1-6667-5701-9 (paperback) | ISBN 978-1-6667-5702-6 (hardcover) | ISBN 978-1-6667-5703-3 (ebook)

Subjects: LCSH: Religion and Science. | Evolution (Biology)—Religious aspects—Christianity. | Naturalism—Religious aspects—Christianity.

Classification: BT712 F50 2023 (print) | BT712 (ebook)

To Joyce,
the love of my life

Table of Contents

List of Illustrations ix

Foreword xi

Acknowledgments xiii

Introduction 1

1. The Origin of Our Species and the End of Christianity? 21

2. The Origin of Our Genus, *Homo* 49

3. The Origin of Our Species, *Homo sapiens* 75

4. The Origin of Our Species and the Ability to Imagine 103

5. The Origin of Our Species and the Cooperative Mindset 132

6. The Origin of Our Species and Providence 165

7. Alpha and Omega: The Beginning and the End 199

Afterword 243

Bibliography 253

List of Illustrations

Figure 2.1. Average performance on four sets of cognitive tests with chimpanzees, orangutans, and toddlers 50

Figure 2.2 Change in Hominin Brain Size Through Time 61

Figure 2.3 Brain Size and Cultural Activity 68

Figure 4.1 Cueva de las Manos 105

Figure 4.2. One of Two Children Found Buried at Sunghir 107

Figure 4.3a and b. Two Views of the Lion-Man. Sculpted out of mammoth tusk 111

Figure 4.4 Chauvet Horses 113

Figure 5.1. Movement of Neural Crest Cells 142

Figure 5.2. Gene 1 and Gene 2 are Master Genes 147

Foreword

THROUGHOUT HISTORY THERE HAVE been times when Christians have felt affronted by the findings of science. Prominent Protestant leaders such as John Calvin and Martin Luther denounced the discovery of Copernicus that the earth moved around the sun and Galileo was accused of heresy. But equally there have been times when the greatest of scientists were people of faith whose research endeavors were sponsored by the church. Gregor Mendel, a man of deep faith,[1] discovered the fundamental principles of genetic inheritance with painstaking research conducted over many years in a monastery garden. We now live in a world that is becoming increasingly polarized across multiple frontiers, with people segregating into different clans based on worldviews of politics, faith, and science. In such an environment it is all too easy to buy into and promulgate the views of the clan without subjecting those views to scrutiny. The borders are hardening and meaningful dialogue between clans is becoming rare. In his first book,[2] Darrel Falk challenged us to think deeply, to speak freely, and to break through faith-science boundaries that need not be there. Here he focuses on what may seem to be the most significant boundary of all, the origin of our own species and how it came into existence through the evolutionary process.

A theology that describes a God who instantaneously created the first human beings has developed over thousands of years. The theological underpinnings of the creation of our species via evolution are not well developed. Falk has taken up the challenge to address questions central to the building of that theology. He has spent many years voraciously reading and deeply contemplating both the scientific literature and Christian theology in order to find answers to pivotal questions. Is the God who has created humans through evolution consistent with the God described in Scripture? If evolution is the creative process by which God brought our species into

1. Richter, "Remembering Mendel," 483.

2. Falk, *Coming to Peace with Science.*

existence, how was God involved and what does it tell us about the nature and purposes God? What emerges is a love story, a picture of a God who loves us and all life on this planet, a God who experiences joy as the process of creation unfolds. Falk describes a growing capacity of our species to think about how this earth came into being and to be social beings capable of reflecting the love that defines God.

In time the church came to grips with the concept that the earth revolved around the sun. A growing body of scientific evidence and a more careful reading of Scripture made it clear that Copernicus was right, that he had pursued truth, discovered and revealed it. Jesus says that he is the way, the truth, and the life. We are called to be truth seekers. Science has revealed a truth about the origins of life on this planet, including that of our species. The acceptance of creation via evolution will grow in faith communities as we work to reconcile this truth with what our Scriptures tell us about God. For those on this journey, or bold enough to start it, Darrel Falk has produced a road map. It has been a labor of love, decades in the making. It is a great scholarly work, meticulously researched, but easily read. This book will make you think. It may make you feel uncomfortable, but we can get a little too comfortable in thinking that we know all that we need to know about God. This is not how strong relationships work; they need to grow to flourish. This book is a beginning of a journey for individuals and churches in gaining a deeper understanding of the God who continues to create and love. Reading it will be an enriching experience.

Philip Batterham, PhD,
Professor Emeritus, School of Biosciences, University of Melbourne,
President, International Genetics Federation,
President (2010–2012), Genetics Society of Australasia.

Acknowledgments

I AM GRATEFUL TO a vast array of individuals who have contributed directly and indirectly to the writing of this book. For almost thirty years, I have been privileged to be a part of a weekly lunchtime science/theology faculty discussion group at Point Loma Nazarene University. To say that my thinking has been shaped by these meetings would be an understatement. The group as a whole read and provided invaluable feedback on the manuscript in the 2021–2022 academic year. I want to thank each person for their participation and for their invaluable feedback on the manuscript. The group included Ron Benefiel, Walter Cho, April Cordero, Rebecca Flietstra, Kerry Fulcher, Rick Kennedy, Mark Maddox, Ken Martin, Stephanie Matthews, Michael McConnell, Michael Lodahl, Denise Nelson, Ross Oakes Mueller, Dawne Page, Dee Puntenney, Heather Ross, Art Seamans, and Dale Shellhamer. I am especially grateful to historian Rick Kennedy for serving as the ongoing organizer and moderator over most of these past three decades. Rick has kept it going and has done a marvelous job. Theologian Michael Lodahl has been an important voice in these meetings, and I am thankful for his influence. Mike read two drafts of this manuscript in their entirety, and even used an early draft as background reading for students in one of his classes. I am privileged to have him as a colleague.

Through that faculty group I became friends with John Wright, a theologian who has guided my reading and influenced my thinking in profound ways. John was only a part of the group in its early days, but I am fairly certain that this book would not have come into existence if John had not sown the seed and nurtured it through our regular meetings over coffee in the years that followed. It was through John's influence that my reading branched out to include great scholars like John Webster in the Reformed tradition, Hans Urs von Balthasar and John Paul II as Roman Catholics, and John Behr in the Orthodox tradition. I am a Wesleyan at heart and John helped me understand that to function in the tradition of John Wesley, one needs to draw heavily from the other great Christian traditions as well.

One of the greatest privileges I've ever had was to participate in the foundational days of BioLogos. The relationships established throughout that time have continued to bless my life in all sorts of ways. I am grateful to Francis Collins for his friendship and counsel during those early days, and to Deb Haarsma for the remarkable job that she and the BioLogos staff have done in building the organization into so much more than we could have ever dreamed in those initial planning sessions. I have been shaped in all sorts of ways—theologically, scientifically, and personally—by my many friends and colleagues in the BioLogos community and beyond.[3] I want so much to name each person but that would change the tone of what I want to say into a long list of names. Instead, speaking to each one—and you know who you are—I want each individual to know how dearly he and she is loved and to sense how grateful I am for our times together.

Many people, other than those already mentioned, have read and provided feedback on this manuscript as it proceeded through its various stages. Specifically, I want to thank Rob Barrett, Arnie Berg, Lyn Cryderman, Russell Edwards, Jeff Hardin, Randy Isaac, Ken Keathley, David Kerk, Richard Middleton, and David Rainey.

The book has been especially enhanced through the Foreword written by the Australian geneticist Phil Batterham. Phil and I were colleagues when I was a young professor in the biology department at Syracuse University and he was a postdoctoral researcher studying evolutionary genetics. He also served as a part-time preacher in various Upstate New York Wesleyan churches and served as a much-loved spiritual mentor to students on our campus. Phil's career has continued to blossom since those early days, as he moved on to become a leader in the field of genetics in Australia and then secretary and president of the International Genetics Federation.

I am especially grateful for the Afterword written by the biological anthropologist Cara Wall-Scheffler and her father, the New Testament scholar and theologian Robert Wall. They agreed to write an Afterword as an imaginary family dinnertime conversation of the kind I am sure they had many

3. I have the used the term "and beyond" because there are many individuals who I've worked alongside who think differently about evolution than we in the BioLogos community. This includes Hugh Ross and Fuz Rana of Reasons to Believe, Todd Wood of the Core Academy and my Southern Baptist friends, especially Ken Keathley. To these individuals I say thanks for those many occasions where our fellowship together showed that we really are one in the body of Christ. I am a better person for loving, and being loved by, each of you.

times over the years. The result is a profound reflection on the central issues I raise in the book. Thank you, Cara and Rob!

Last, but most certainly not least, I want to thank my family for their abiding love and support. My brother, Doug, has been an ongoing encouragement—driving through snow and ice and even flying many miles to attend my talks on occasion. It was always a huge encouragement to look into the audience and see that one familiar face, the person who shared a bedroom with me for almost two decades, and to sense his never-dying love and support. My daughters, Cheryl and Shelley, and their husbands, Damian and Murray, and my grandchildren, Caleb, Noah, and Sara, have continually been the inspiration that has refreshed my soul and nourished my spirit. Your presence in my life has played no small role in making this book possible. Finally, the single most influential and manifestly important component to anything that I've ever done that has been worthwhile in life is the supporting love of my wife, Joyce. This book is dedicated to her.

Introduction

> There is grandeur in this view of life, with its several powers, having been originally breathed into a few forms or into one; and that, whilst this planet has gone cycling on according to the fixed law of gravity, from so simple a beginning endless forms most beautiful and most wonderful have been, and are being, evolved.
>
> —Charles Darwin, 1859, *On the Origin of Species*

Endless forms most beautiful, indeed, but in his ground-shaking, paradigm-shifting publication, *On the Origin of Species,* Charles Darwin chose not to discuss the most amazing of all species—humans and their incomparable mind. Having mulled over the consequences of presenting his theory for more than twenty years, when he finally laid it out, his mention of the origin of humans was reduced to a single sentence: "Light will be thrown on the origin of man and his history." Darwin had an extensive set of notes on the origin of humans and was prepared to write a whole chapter on this topic, but in the end he chose to set it aside. Why did he essentially leave humans out of his story—surely this was the most important question of all? As Darwin historian Janet Browne put it: "He knew that anything he said was bound to ignite furious controversy, and anticipating just such a response, he . . . drained his manuscripts of any reference to a Creator or human ancestry."[1]

Sure enough, despite his effort to sidestep the issue, this quickly became the matter that took center stage. As Darwin's *Origin* was coming off the press, the British public was just becoming aware of the existence of African gorillas. Darwin's theory conjured up images of an ancestral lineage that was emotionally disturbing and repugnant to many. The clergy at

1. Browne, *Power of Place,* 60.

Cambridge reportedly became concerned that "a bestial evolution would cause a loss of faith."[2] Samuel Wilberforce, bishop of Oxford, participated in a famous exchange about seven months after the publication of *Origin*. Emotions ran high as the exchange between him and the Darwin defender, Thomas Huxley, began. At the debate, held at Oxford, many students were in attendance and a chant of "Mawnkey! Mawnkey!" raised the fever of the crowd.[3] This was no ordinary scientific discussion. Browne writes,

> Perhaps it was laughter that loosened Wilberforce's grip. At one point he turned to Huxley . . . Was [he] related on his grandfather's or grandmother's side to an ape? . . . The gibe was understood by every member of the audience. They smelled blood. So did Huxley. "The Lord hath delivered him into mine hands,"[4] he apparently whispered to Benjamin Brodie on the bench beside him.[5]

And, as Browne describes it, with that Huxley apparently stood up and proceeded with his famous reply:

> If I would rather have a miserable ape for a grandfather or a man highly endowed by nature and possessed of great means and influence, and yet who employs those faculties for the mere purpose of introducing ridicule into a grave scientific discussion—I unhesitatingly affirm my preference for the ape.[6]

Meanwhile, Robert Fitzroy, the former captain of the *Beagle*, for whom Darwin had served as companion and naturalist in their voyage around the world twenty-five years earlier, was in the crowd. Waving a Bible, Fitzroy implored anyone who would listen to believe God's Holy Word rather than that of a mere human on the question of creation.[7] So even though Darwin had done his best to keep the volatile question of human origins out of his book, it didn't take long for it to rise to the surface as the most controversial issue of the day.

2. Desmond and Moore, *Descent of Man*, xxxv.

3. Parts of this account may be apocryphal. If so, we will never know which parts. I have relayed it here as recounted by Browne, *Power of Place*.

4. This is a reference to 1 Samuel 23:7, which in the KJV states: "And it was told Saul that David was come to Keilah. And Saul said, God hath delivered him into mine hand; for he is shut in, by entering into a town that hath gates and bars."

5. Browne, *Power of Place*, 122.

6. Browne, *Power of Place*, 122.

7. Browne, *Power of Place*, 123.

I begin this book on human origins with that story because in some regards, 160 years later, it seems nothing has changed. In a 2019 Gallup Poll, 40 percent of Americans questioned believed "God created man in present form," a percentage that is almost the same as when this poll was first conducted almost fifty years earlier, at which time it was 44 percent.[8] How could it be that a theory which in broad outline has been verified in thousands of experiments over many decades and is foundational to the entire discipline of biology, nonetheless be widely rejected by the general public? There is something disconcerting about human evolution that makes it likely the hardest of all scientific concepts to reconcile with day-to-day living. For various reasons, this is especially true for the Christian. Darwin thought his theory closed the door to the reasonableness of Christian faith and far too many have feared that he was right. Emphatically, this book seeks to show that is not the case. Instead, I hope to show that what we are learning scientifically about the specifics of how our species evolved into its current state opens the door to a vista of wide-open, largely unexplored spaces that enrich our understanding of what it means to have been lovingly and purposefully created by God.

This book was born out of my own set of personal experiences. I grew up deeply immersed in a church community that placed a great deal of emphasis on living a holy life and doing so within the context of a rich and meaningful personal relationship with God. In my small church, there was a tradition for congregation members to select their favorite song, which we would then sing together. I am sure that little congregation of about forty members grew quite tired of singing my favorite because I asked for it over and over again. Each verse started with the line, "Jesus is all the world to me," and concluded with the simple but life-affirming sentiment, "He's my friend." Those words depicted the peace-filled world in which I lived almost all the time. Almost! There was one especially perduring problem from which I could not escape. I had an ongoing fear that scientists had disproved the Bible's origin story, and I went through periods of excruciatingly painful doubt. I was deeply concerned that all those smart people—the scientists—might be right. We were just people of average intelligence; I reasoned, how could I be sure we were right and they were wrong? Being only a child and then a teenager, I certainly did not know enough to critique their analysis.

8. Gallup, "40 percent."

So, mostly I just tried not to think about it and Jesus remained my best (and hopefully not imaginary) friend.

It is now many years later, and I have come to know many biologists who are devout followers of Jesus and would testify to the harmony between their faith and science. Indeed, I have become a biologist and my understanding of the magnificent beauty of biological processes only deepens my faith. This is especially true of the beauty of what we are learning about how our species came to be. Still, however, surveys show that almost all leading biologists (as defined by membership in the National Academy of Sciences in the United States or the Royal Academy of Arts and Sciences in Britain) do not believe in a personal God (~95 percent).[9,10] I suppose I've read hundreds of biology books, many of them on human evolution, and I have seen few, if any statements in those secular books that lead the reader to believe that the possibility of the personal God of orthodox Christianity is still on the table. This contrasts with the scores of statements I have seen that lead the reader to believe that science has shown that an orthodox Christian perspective is no longer viable. These authors all seem to assume that any God that might exist has been completely removed from any aspect of the biological world for billions of years. In an especially noteworthy book, the late Stephen Jay Gould defined science and religion as occupying separate, nonoverlapping magisteria. Religion occupies a separate sphere from that studied by science and it is confined to questions related to how we should live. His foremost rule within this worldview is that God does not "directly ordain important events in the history of nature by special interference."[11] This, as I see it, is the assumption in many classroom biology textbooks. As I will show in pages that follow, this is a worldview assumption, grounded not in scientific data, but based on a particular philosophical viewpoint—one that permeates the evolutionary science community. Students are led to believe (directly or by inference) that science has now shown that the existence of divine activity has played no direct ongoing role in the creation of our species. So, just as I felt decades ago, a young person can easily be under the impression that a person must choose between the science represented by evolutionary biology and the Christian faith. It is true that the Christian

9. Larson and Witham, "Leading scientists still reject God."

10. Stirrat and Cornwell, "Eminent scientists reject the supernatural." For a thorough analysis of scientists in general at top-tier universities, see Ecklund, *Science vs. Religion*. In this study, only 2 percent of scientists in general considered themselves evangelical Christians, but the book carefully considers various nuances. Nine percent, for example, indicated that they had no doubt about God's existence, while another 14 percent indicated that although they sometimes had doubts, they do believe in God.

11. Gould, *Rocks of Ages*, 84, 85.

community now lays out many books and has whole organizations with an alternative narrative, but most of them are grounded in a notion that unnecessarily pits Christianity against core facts in biology.

For the first third of my career, I was a geneticist immersed in the secular research community. I attained tenure at Syracuse University and fully expected to spend my whole career there. But I chose to move into the Christian university setting in part because I was deeply concerned that Christianity was unwisely seeking to put forth a new science, and one that my community considered to be noteworthy for its poor, easy-to-dismiss quality. That was not what was needed. It was the worldview of scientists like Gould that was flawed, not his evolutionary biology. So, for the past two-thirds of my career, I have been on the biology faculty of one or the other of two Christian universities. From the day I was interviewed at each of the two, I tried to make one thing clear. Even though a significant proportion of my students would be coming into the institution from churches and schools where evolution is considered scientifically flawed, as a biologist I strongly disagreed and would not shy away from explaining why. I was convinced that the evidence for evolution was overwhelming, and I would not support the teaching of a "scientific" perspective outside of the peer-reviewed scientific mainstream. About nine years into my time at Point Loma Nazarene University (the fall of 1997), our provost called and asked me if I would talk with a person who had a concern about how we teach biology. This person had connections to someone with significant resources, so the matter was sensitive—especially in a tuition-driven institution like ours. The provost was totally supportive of my position but felt that it was important that the rationale be explained. So that was my task—please explain what we do and why. I quickly learned that this outsider wanted to arrange for a course in young earth creationism on our campus. It would be taught *pro bono* by a "highly qualified" person with a graduate degree and connections to a young-earth organization across town. I arranged several visits with this man, during which time I was hoping that perhaps he, as a nonbiologist, might get a glimpse of why we in biology would not be able to have a course like that taught under the context that it was credible science. When I finally put my foot down and said no, his response was predictable. As he stormed out of my office, he told me that there would be a tsunami coming my way and that it would sweep me and the reputation of our university as a Christian institution into oblivion. His concern harkened back to events a century and a half earlier on another university campus. Accepting Darwin's theory of a "bestial evolution" would lead to a loss of Christian faith. The chant of those Oxford students—"Mawn-key! Mawn-key!"—could almost be heard echoing in the hallways as I watched him leave.

At the time, I knew of no book written by a biologist for a conservative Christian audience that I could give to this man explaining the logic of evolution in a manner that would show why it is fully consistent with a Christian faith firmly grounded in the Bible as the divinely inspired word of God. He needed to be able to see the rationale for why almost all biologists think that evolution has occurred, but he needed to see it in the context of how a person's faith in the life, death, and resurrection of Jesus is not impoverished but is actually enriched by this reality. Because of that, I set out to write the manuscript that eventually became *Coming to Peace with Science*, my first book. As my writing proceeded, a draft got into the hands of someone with connections to my tsunami visitor. Apparently, it didn't help offset any of his concerns. Indeed, I'm pretty sure he thought to himself: "The Lord has delivered him into mine hands," and with that began an onslaught that continued for the next two and a half years. Far from soothing any of his fears, my book provided him and others with the ammunition needed to foment a full-scale attack. Sure enough, the efforts came in waves, each of which was designed to knock me out of my faculty position and rescue the institution from faculty members like me. These challenges would have been personally overwhelming were it not for the strong support of church and university administrators who stood firmly behind the principle of academic integrity despite the successful attempt to draw in some of the most influential persons in evangelical Christianity. Given that Christian universities are tuition-driven, there was much at stake financially if our institution received the negative publicity that was threatened. At one point, I was sent a letter containing a set of false statements that I was told would be attributed to me and publicized widely unless I denied them within a tightly prescribed time limit. I did. At another point, I was told by a leader of the Intelligent Design movement that a major Christian radio program was committed to doing a program to show that faculty at Christian colleges were "more interested in accommodating to the naturalistic worldview than with challenging it." This person had written a book showing (in his opinion as a legal scholar) that evolutionary theory was deeply flawed; he wanted Christian colleges to challenge this theory according to his formula. He was not a biologist, and I thought his analysis reflected that. The clear implication was that the radio show to which many of the parents of prospective students listened would focus on my university, with me as the prime example of what we were not doing right. We had always tried to be forthright about our perspective on the evolution question, but this was scary to us all because given the twists that they would put on the story it had the potential to do massive reputational damage. A small team including the chair of our biology department, a member of our university's administrative cabinet, and I traveled to the headquarters of the organization responsible for the radio program. We also had supportive letters from twenty-eight former students who attested to the faith-strengthening value of their experience in the biology department. Fortunately, there was a spirit of God's love permeating the meeting, and although we disagreed on how God created, any plans for a reputation-damaging radio program were shuttered.

Finally, after almost two and a half years of these sorts of experiences, the last and potentially most consequential episode was my requested appearance before the governing board of the university. The board was respectful, but they wanted to hear a summary of what we taught in the biology department, and how it impacted the Christian faith of our students. There was much on the line, but I will never forget the peace that came over me on the Sunday morning before the meeting, as the Bible study group I led at my church supportively gathered around. Placing their hands on my shoulders, they prayed for me and the defense I would be mounting before the board. The meeting took place five days later. When it was my turn to speak, I told the story of my own life-wrenching struggle with how to put the theory of evolution into the context of my Christian faith. It included a discussion of how supportive a particular church had been to me and my family as I tried to find my place as both a biologist and a person committed to following Jesus.[12] After I explained why I thought it was important to guide biology students through the basic evolutionary principles in a Christian context to prepare them for what would lay ahead, the board members stood and warmly applauded. With that, I felt relieved—a standing ovation no less. However, a bit like Darwin I suppose, I had chosen not to discuss my conviction that humans had arisen through the evolutionary process as well. So, early during the question period, one of the board members described a visit he had taken to the Smithsonian National Museum of Natural History in Washington, DC, where the "supposed" lineage to humans was presented in vivid detail. "What do you think of that?" he asked me kindly. I responded by telling him that I thought this accurately reflected our ancestral lineage. "This is how God has created humankind, as well," I said. With that, the mood changed dramatically and there was a collective, almost audible gasp in the room. Indeed, I was quite certain that many of the board members would have gladly taken back that standing ovation of several minutes earlier if only they could have. In the minds of many, it was one thing to accept that animals arose from more primitive forms of life, but a whole other matter to accept that this is our human heritage too. After I was asked to leave for the closed-door session, there was a frank discussion,

12. That personal struggle is discussed in my book, *Coming to Peace with Science.*

I was later told. For me, the standing ovation became a faded memory compared to the impression of those concerned looks on board members' faces at the seemingly heretical notion that we humans had descended from apes.

In the end, I survived; the threatened tsunami never arrived, and I was able to continue as a biology professor at my beloved institution until moving into emeritus status many years later.

As mentioned, the manuscript that provided the energy for those tsunami attempts ended up becoming my book *Coming to Peace with Science.* Like the talk I gave before the board, and indeed like Darwin's *Origin of Species,* I chose to say almost nothing about human evolution in that book. In fact, knowing that the word "evolution" was synonymous with a godless view of origins in the mind of my intended audience, I decided not to use the word "evolution." Instead, I chose to call it "gradual creation" to emphasize that the process was God's means of creation, and not an impersonal algorithm. It is now two decades later and still, to my knowledge, there is no broad-ranging, theologically oriented book by a Christian biologist that builds upon the mounting scientific data informing us specifically about how God brought our species—unique as it is—into being. The data indicating that our species was created through the process of common descent are well known and will be summarized in chapter 2 of this book, but the key question remains: How did this come about through divine activity? If God created our species through the evolutionary process, how ought we think about this both scientifically and theologically? There is a growing consensus among scientists over the last couple of decades about how our unique human qualities have come about through the evolutionary process, but this understanding needs to be brought into a theological context. The scientific data are summarized in provocative book titles like *Survival of the Friendliest, The Goodness Paradox, The Secret of Our Success, How Compassion Made Us Human, The Creative Spark,* and *Super Cooperators: Altruism, Evolution, and Why We Need Each Other to Survive,* by scholars at leading universities like Duke, Harvard, York, and Princeton. These are science books, though, and they do not attempt to address the possibility of how God, if the divine even exists, may have played a role in bringing about the "friendliness," "goodness," "compassion," "creativity," and overall "success" of our species. The fact is that there has been a set of harmonizing advances in genomics, evolutionary genetics, paleontology, anthropology, and archaeology that are making it increasingly clear just how compatible the science of human origins is with core aspects of the Christian faith. The purpose of this book is to show just how theologically fulfilling the scientific advances are turning out to be.

Science is a gift from God to reveal truths about how creation operates. It can be abused, but rather than fear science, it is important for us to recognize that it facilitates our praise as we more knowingly celebrate the beauty of the natural world. Recently discovered details about the origin of our species are no exception. Science, correctly understood, and Scripture, rightly interpreted, exist in a worshipful, God-ordained, synergistic relationship.[13] Science also facilitates our divine calling to live into the image of God, to be God's representatives on this earth. As we grow in our understanding of the way in which creation operates, we can participate in building a better world—one centered in God's care for all people. In this scientific age, where nontheistic interpretations of the origin of our species abound, it is important to shift those findings from their portrayal that our appearance on earth is a rudderless cosmic accident. This is untrue and the propagation of this narrative will have devastating consequences. Instead, our origin as a species is tied to our vocation which—by imaging God—is to represent the loving care of God for each other and for all of creation. This is the narrative that lies at the center of this book.

One of the premises of this book is that we have been created out of, and brought into, the love of God. God's love did not begin with Abraham, Sarah, and their descendants. God, by definition, is love and that love has no beginning.[14] I will develop the thesis that God's love was being manifest on those African savannahs over 100,000 years ago regardless of how well it would have been understood. This is consistent with the thinking of the great Christian revivalist of the eighteenth century, John Wesley. My colleague Michael Lodahl summarizes this:

> For Wesley, there is no human being who has ever existed who was not deeply loved and immediately graced by God . . . Thus, no human being lives without the Spirit's wooing to live compassionately and justly with (all) others. . . . [Wesley wrote,] "Everyone has some measure of that light, some faint glimmering ray, which sooner or later, more or less, enlightens every man that cometh into the world."[15]

The Bible's first eleven chapters starts with a discussion of human origins in the context of a relational God. It then goes on to describe the human predicament showing that God is always seeking to restore relationship despite human sinfulness. Biblical scholar and theologian Terence E.

13. The organization BioLogos exists to demonstrate this principle. Its website, BioLogos.org, is a marvelous resource.

14. "God is love. Whoever lives in love, lives in God, and God in them" (1 John 4:16).

15. Lodahl, *Matthew Matters*, 82.

Fretheim emphasizes the importance of recognizing the ramifications of this biblical focus:

> This relational God of the Old Testament is not, first and foremost, the God of Israel but of the world. The opening chapters of Genesis, for example, make universal claims for this God. These Genesis chapters portray a God whose universal activity includes creating, grieving, judging, saving, electing, promising, blessing, covenant making, and law giving. And we are not yet to Israel! God was in relationship to the world before there ever was an Israel, and so God's relationship with Israel must be understood as a subset within this more inclusive and comprehensive divine-world relationship. God's acting and speaking are especially focused in Israel, but this divine activity is a strategic, purposive move for the sake of the world (Gen 12:3, "in you all the families [the "families" in Genesis 10] of the earth shall be blessed"). The election of the family of Abraham and Sarah is an initially exclusive move for the sake of a maximally inclusive end.[16]

Similarly, in the New Testament, the most theological of the four Gospels, John, begins by placing the life of Jesus into the context of the creation of *all* humankind. John 1:4 states, "In Him was life and that life was the light of *all* humankind." And the ninth verse goes on to say, "The true light, which enlightens *everyone*, was coming into the world." That light, John is telling us, did not begin to shine with the birth of Jesus. That light, the second person of the Trinity, is eternal. Like God's love, there was never a time or a place in which the light of God did not shine. It was often ignored and even darkened by the force of evil, but it was always there.[17] As Jesus said, "Before Abraham was born, I am."[18] If those people living on the savannahs and shores of Africa over 100,000 years ago had the same sort of needs and desires that we do, then would not the love God has for all people have been present in the midst of their daily life, especially when they struggled? Furthermore, would there not have been periods of time when (or even particular people through whom) the light of God would have been able to break through? We will explore these questions more carefully in the coming pages.

That is the overall background behind this book. To summarize:

16. Fretheim, *God and World*, 18–19.

17. "The light shines in the darkness, and the darkness has not overcome it" (John 1:5).

18. John 8:58b.

- There has been a long running disconnect between the science of biology and biblically grounded Christianity stretching back more than 160 years.
- In part, the disconnect is a result of the worldview of the leading spokespersons of the evolutionary sciences.
- In part, the disconnect is the result of conservative Christians trying to pick apart the science of evolution, when the real issue is the unwarranted philosophical convictions of many of its practitioners.
- In part, the disconnect is the result of conservative Christians not yet theologically addressing the nature of divine activity in human creation in a manner that is informed by the scientific analysis of human evolution.
- This book seeks to build a bridge across the 160-year-old gap, by showing how human evolutionary science is thoroughly consistent with orthodox Christian theology.

So that is the book's purpose. How is the book structured as it sets out to accomplish this task? It is to that matter we now turn.

Chapter 1, *The Origin of Our Species and the End of Christianity?*, begins by addressing the myth that has been propagated by many of biology's key spokespersons. Indeed, many of my scientific heroes—those who have done fantastic work in showing how living systems function—Francis Crick, James Watson, Jacques Monod, Richard Lewontin, Edward Wilson, Richard Dawkins, and Stephen Jay Gould—are atheists or, at least in the case of Gould, effectively atheistic. Francis Crick, for example, became an atheist at the age of twelve and he spoke freely about the fact that his atheism was a factor motivating his scientific agenda.[19] In *The Eighth Day of Creation*, historian Horace Judson wrote, "An important reason Crick changed to biology, he said to me, was that he is an atheist, and was impatient to throw light into the remaining shadowy sanctuaries of vitalist illusions."[20] Similarly, in his autobiography, Crick wrote, "What would be more important than to find our true place in the universe by removing one by one these unfortunate vestiges of earlier beliefs?"[21] This sort of writing extends into the popular

19. Crick, *What Mad Pursuit*, 10.

20. Judson, *Eighth Day*, 87.

21. Crick, *What Mad Pursuit*, 11.

scientific literature. For example, the noted author David Quammen writes: "What [evolution theory] challenges is the supposed godliness of Man—the conviction that we above all other life forms are spiritually elevated, divinely favored, possessed of an immaterial and immortal essence . . ."[22] The science associated with evolutionary theory is not capable of addressing spiritual questions, such as whether humans are in some sense immortal. What sort of scientific experiment could address a question like that? So if Quammen is right and evolutionary theory really does "challenge the supposed godliness of Man," then he has moved this aspect of evolutionary theory out of the scientific realm and into the domain of religious philosophy. Sadly though, he doesn't tell his readers what he has done, and his wording implies it is science itself that has pushed the concept of humans having been created in the image of God out of the picture. These subtle switches—back and forth between worldview and science—have made it difficult for the general public to sort through what is a real scientific conclusion, as opposed to one that is really a philosophical assumption. So, what are the limits of science and when are biology's spokespersons crossing those lines? That is one all-important matter that is the focus of chapter 1.

But it is not enough just to point out the messiness brought about by conflating scientific and worldview conclusions. The question still remains: since science has been enormously successful at explaining natural processes without needing to invoke a supernatural presence, is the worldview that the supernatural really exists still reasonable? Does any well-founded logic remain for thinking the divine exists or is it just wishful thinking? The apostle Paul once wrote that if Christ was not raised from the dead, Christians "are of all persons most to be pitied."[23] The atheist evolutionary geneticist Jerry Coyne latched onto these words of Paul, claiming correctly that the supposed occurrence of the resurrection is the linchpin of Christian theology.[24] Coyne, of course, thinks that belief in the resurrection is a childish fantasy. But is it? This is the crux because if the resurrection of Jesus Christ really did occur, it would mean there is more to reality than what we are able to see with our eyes, touch with our fingers, and measure with our scientific instruments. Is there any way of examining the reasonableness of this belief or will it always be only in the realm of what is hoped to have been true? Chapter 1 examines this carefully through the work of a set of historical scholars and concludes that not only is it a highly reasonable historical hypothesis, but there is reason to think it is the *most* reasonable explanation.

22. Quammen, *Reluctant Mr. Darwin*, 209.

23. 1 Cor 15:19b.

24. Coyne, *Faith versus Fact*, 44.

The chapter goes on to show that it *is* logically possible to build one's life around the supposition that we are a purposefully created species, and not an accidental result of an impersonal algorithm.

From there the book turns to what the scientific investigations *do* tell us about how our species came into existence. This is done in two phases. Chapter 2 focuses on the origin of our genus, *Homo,* and chapter 3, on the origin of our species, *Homo sapiens.* The narrative begins with a discussion of why we can be quite certain that we did indeed share a common ancestor with chimpanzees and bonobos by examining genetic and fossil evidence. The story then follows the gradual change in the fossils of our family tree as its members came to acquire the distinctive key features that characterize our lineage. Increasing brain size is an especially noteworthy feature, and it began to distinguish our ancestors about two million years ago. The genetic changes that likely played a key role in setting this brain alteration in motion are fascinating and, even though this is a book for nonspecialists, I think the changes can be explained in a way that gives the reader a sense of excitement for what we are learning about how these changes took place at the biological level. Following the chapter 2 overview, we are ready to explore a couple of theological questions that emerge as part of the context of a long, multi-million-year journey to humanness. What, for example, does this enormous length of time imply about God as Creator? Why so long? Could it be, chapter 2 asks, that God delights in the entire long parade of creation through time? After all, Genesis 1 tells us at the end of each stage that "God saw that it was good," and John 1 informs us that "without him nothing was made that has been made." Creatures—all creatures—have always been embedded in the love that flows from God into creation.[25] It is not all about us.

When our species gradually emerged in Africa about 200,000 to 300,000 years ago there were still at least seven other *Homo* species in Africa, Europe, and Asia. Many of these species still existed as some members of our species left Africa and entered the Middle East 70,000 or so years ago. However, by 39,000 years ago, all the others had gone extinct. Up until that point, for several million years, there had been multiple hominin[26] species, sometimes even in the same geographical area. As described in chapter 3, many investigators think that something had happened that had given our species a distinct adaptive advantage, and the question for the ages has been what that quality (or set of qualities) was. There is a growing consensus, though, that what set *Homo sapiens* individuals apart was their highly

25. See Ps 104, for example.

26. The term "hominin" refers to those beings distinctively on the lineage to human beings, but not on the lineage to chimpanzees, bonobos, or gorillas.

sophisticated, "others-focused," social brains. We, like no other species, have unique skills in understanding the minds of others, and through that quality, we are inimitably capable of working together to accomplish mutually desirable objectives. The key question examined throughout chapter 3 is how this quality came about. The ability to understand that another individual has a mind like one's own and to be able to form a "theory" about what an individual is thinking is called "theory of mind." Chimpanzees have a rudimentary theory of mind—they can sometimes understand the intentions of another and can gauge their own behavior based on their perceptions of the intentions of another. However, our species takes this to a whole different level and our entire existence is tied to our unique ability to sensitively understand the emotions, intentions, and perceptions of others as though they were our own.

It is here that we come to the key point that is the focus of this book. Once the ability of the hominin mind became sophisticated enough for individuals to understand that others had a mind very much like their own, those minds would have been able to understand not only each other in a more meaningful way, but they would have been able to understand the promptings of God's Spirit in a more meaningful way as well. Since God, by definition, is love, God's love could have been experienced in an increasingly nuanced fashion as that theory of mind was becoming more and more sophisticated. Indeed, could it be that what we see in the development of our children today as they proceed from the highly self-focused mind of a baby to a six-year-old's increased awareness of what is on the mind of others parallels in some sketchy fashion what happened to members of our species as their theory of mind was emerging in evolutionary history? The promptings from God, grounded as they would have been in love, might sometimes have influenced their behavior just as Christians believe those promptings influence our behavior today. Chapter 3 describes the ramifications of this Spirit-based influence by examining it through the lenses of evolutionary biology, genetics, and (significantly for this book's purpose) theology.

Chapter 4 is called "The Origin of Our Species and the Ability to Imagine." Here we will examine the question of whether there is evidence for a divine-human interaction of the sort that we might expect if there was a population that was especially paying attention to what is sometimes called the wooing of God's Spirit. Have archaeologists identified sites where there is an indication of a changed mindset—one illustrating that people were not just becoming smarter, but that they had a mindset specifically focused on others? Further, is there archaeological evidence that certain communities may have been developing a sense that there was something outside of their present reality—a life beyond? Of course, we can expect there would be secular explanations for these sorts of artifacts, but might they, at least in some cases, be the result of an enhanced communication with the divine?

Chapter 5 is entitled "The Origin of Our Species and the Cooperative Mindset." Intriguingly it was Charles Darwin who first recognized that compassion may well have been an important driving force in our success as a species. In his 1871 book *The Descent of Man*, Darwin suggested that if humans had sprung from a formidable, large, ferocious ape, our ancestors would not have had to depend on each other in the way we do. If that had been our heritage, he suggested, it would have "checked the acquirement of the higher mental qualities such as sympathy and love for his fellows."[27] So Darwin did not think it likely that we came from such ferocious beings. Instead, as he put it, there would have been an immense advantage for "man to have sprung" not out of strength but out of "comparative weakness." In other words, Darwin thought that the secret to our success as a species may have been that we needed each other and the qualities that enabled us to thrive best were mental qualities that facilitated care and understanding:

> After the power of language had been acquired, and the wishes of the community could be expressed, the common opinion how each member ought to act for the public good, would naturally become in a paramount degree the guide to action. But it should be borne in mind that however great weight we may attribute to public opinion, our regard for the approbation and disapprobation of our fellows depends on sympathy, which, as we shall see, forms an essential part of the social instinct, and is indeed its foundation-stone.[28]

Darwin went on to stress that on a long-term basis "habit" would play a key role in "guiding the conduct of each member" of the community. The word "habit" for Darwin did not mean what it means for us today. Genes had not yet been discovered, but that is what Darwin meant by that word. It becomes clear from the context that he was speaking of an inherited behavior, one which today we would say was embedded into our genetic makeup. Darwin went on to discuss the significance of empathy in a manner that would depend upon each individual recognizing that others had a mind like their own.[29] Indeed, he even wrote that those ancient little communities would have had a sense of right and wrong and suggests that this would have been encoded in what we today call genes.

27. Darwin, *Descent of Man*, 84.

28. Darwin, *Descent of Man*, 122.

29. Darwin, *Descent of Man*, 129.

As was so often the case, Darwin was well ahead of his time. But modern science has been catching up with him and the question we can pose now is just what are the genetic changes associated with what Darwin called "habit"? And that is the topic of chapter 5.

Chapter 6 is called "The Origin of Our Species and Providence." There were two issues that greatly concerned Darwin as he rejected the faith of his youth. The first is that he was deeply steeped in the thinking of William Paley, who was one of the most influential theologians in the Church of England at the time. Darwin, as a student at Cambridge University, was required to read Paley's *Evidences of Christianity* and his *Moral and Political Philosophy*. In addition, on his own, Darwin read and thoroughly studied Paley's *Natural Theology*. Darwin biographers Desmond and Moore write, "Paley's logic so delighted Darwin that he learnt it by heart. The sheer elegance of the deductions fascinated him: given that God exists, he would obviously reveal himself, and how else but by miracles."[30] However, as time went by Darwin concluded that this view of God as the master designer did not fit with his theory of natural selection. In his autobiography, Darwin wrote, "We can no longer argue that, for instance, the beautiful hinge of a bivalve shell must have been made by an intelligent being like the hinge on a door by man."[31]

The second issue that contributed to his loss of faith was that he considered the massive suffering of animals associated with natural selection to be inconsistent with the God of the Christian faith. This, together with human suffering, including that of his beloved daughter, Annie, who died at ten years of age, put an end to the faith of his youth. His great-grandson Randall Keynes writes,

> After Annie's death, Charles set the Christian faith firmly behind him. He did not attend church services with the family; he walked with them to the church door, but left them to enter on their own . . . While others had faith in God's ultimate goodness, Charles found him a shadowy, inscrutable, ruthless figure.[32]

Each of these concerns Darwin had (and many of us have as well) are examined in chapter 6. Interestingly, even though Darwin came to reject the traditional Christian faith as he understood it, his views did not rule out a divine purpose of some sort. The process of evolution took place within the context of a world that operated by natural laws and he took great pains to

30. Desmond and Moore, *Life of a Tormented Evolutionist*, 78.

31. Darwin, *Autobiography*, 87.

32. Keynes, *Annie's Box*, 222.

show that natural selection itself was lawlike. For example, in *On the Origin of Species,* Darwin wrote the following:

> To my mind it accords better with what we know of the laws impressed on matter by the Creator, that the production and extinction of the past and present inhabitants of the world should have been due to secondary causes, like those determining the birth and death of the individual.[33]

Indeed as R. J. Berry wrote in his book chapter, "Did Darwin Dethrone Humankind?," Darwin retained a respect for the possibility of God's existence to the very end:

> One thing can be asserted strongly: Darwin never strongly renounced his faith. Towards the end of his life, he replied to the atheist John Fordyce who asked if belief in evolution and God were compatible, that "It seems to me absurd to doubt that a man may be an ardent Theist and an evolutionist . . . as to my own views . . . in my most extreme fluctuations I have never been an atheist in the sense of denying the existence of a God." One of his last letters . . . declared "my inward conviction is that the universe is not the result of chance."[34]

Thus Darwin, despite his doubts, left open the possibility of divine purpose in the process of human creation. He stated, "I cannot look at the universe as blind chance . . ." He clearly thought of natural selection as a teleological (purpose-driven) process, one leading to highly adaptable living systems. But he was also convinced that randomness played a significant role in the appearance of the sorts of variants upon which the process of natural selection depended. There was nothing tying the potential usefulness of a variant to its production. Chance alone dictated whether a useful variant would appear. But given enough time and a large population of organisms the process of natural selection was very predictable. It would lead to organisms that are highly adapted to survival in particular environments.

The fact is that Darwin rejected the faith of his youth in no small part because of an inadequate view of providence—one deeply steeped in the religious thinking of nineteenth-century England—but he never became an atheist. It is true that Darwin stressed the significance of chance in generating variations, but he nonetheless put the emphasis on the creative power of natural selection and the profundity of its existence likely played a key role in his holding onto the likeliness of some form of transcendent reality.

33. Darwin, *Origin of Species,* 490.

34. Berry, "Did Darwin Dethrone Humankind?" 37–38.

However, a century later, in 1971, in a book called *Chance and Necessity,* the molecular biologist Jacque Monod shifted the focus and randomness took center stage. Monod understood the power of natural selection in producing highly adaptable life-forms, but he put the emphasis as to which was more important in a different place. Monod wrote that "Man was the product of an incalculable number of fortuitous events, the result of a huge Monte Carlo game . . ." Fifty years later, the highly influential biologist Sean B. Carroll essentially pits the emphases of these two individuals, among the greatest biologists of the last two centuries, against each other:

> Let's dare to ask which of the two bona fide geniuses is right? Or more right? The answer matters because the greater the creative role of random mutation, the more life appears per Monod to be driven by chance.[35]

Carroll argues for the importance of chance[36] in a manner that I describe more fully in chapter 6; he concludes that Monod's emphasis is the correct one:

> The domain of chance has expanded into realms [no one ever] imagined. As we have learned much more about the history and workings of the planet, we have been startled to discover how the course of life has been buffeted by a variety of cosmological and geological accidents—without which we would not be here.[37]

If he (and Monod) are right, then the lawlike activity of natural selection, although certainly important, is not the chief driving force influencing which specific life-forms, including our own, appear on earth. Despite the fact that Darwin's natural-selection algorithm is the engine that powered the process that led to *Homo sapiens*, the actual pathway taken was an extremely improbable one. There was a vast array of possible pathways, only one of which would have finally led to us. Our existence as a species, Monod tells us, was sparked by one chance event after another, any one of which if it had not occurred would have led to different outcomes. Fifty years ago, Monod confidently asserted that "nothing warrants the supposition—or the hope that on this score our position will ever likely be revised." Then he

35. Carroll, *Series of Fortunate Events*, 100.

36. It is important to define what we mean by chance in this discussion. We are using the term in the same way that Sean Carroll does. He writes, "By chance I mean a rare, unpredictable, or random event, or one that entails so many variables or forces that it is near enough to be random." Carroll, *Series of Fortunate Events*, 59.

37. Carroll, *Series of Fortunate Events*, 10.

goes on to write, "There is no scientific concept, in any of the sciences that is more destructive of anthropocentrism than this one."[38] Carroll concludes that not only was Monod right, the case has become much more solid than he could have ever predicted fifty years ago. In chapter 6, we show that this view is now the mainstream view in biology, and we examine the implications of this for how we think about the providence of God in bringing our species into existence.

All of this leads into the book's final chapter (chapter 7), with its underlying question: How do we think about the biblical creation narrative itself from beginning to end? And, given the emotions this question has been triggering for the past 160 years, how do we have conversations about this in a manner that is always grounded in the single most important tenet of the Christian faith? Paul writes, "And now these three remain: faith, hope, and love. But the greatest of these is love" (1 Cor 13:13). Einstein, according to one apocryphal musing, said, "I want to know the thoughts of God, the rest are mere details." As we move into the final chapter of the book entitled "Alpha and Omega: The Beginning and the End," we are drawn into the fact that "knowing the thoughts of God" entails becoming immersed in God's love. If God by definition is love (1 John 4:16), then as we know and live into love—for each other and not just God—it is there that we are truly able to come to understand what is at the heart of how our species came to be.

38. Monod, *Chance and Necessity*, 112.

1

The Origin of Our Species and the End of Christianity?

THE LIMITS OF SCIENTIFIC KNOWLEDGE

AMIDST THE HUSTLE AND bustle of being a new professor of biology at Syracuse University in the mid-1970s, one of my favorite respites was Saturday mornings in the library at the medical school. There, I would glance through the recent journals to explore the latest findings in the exploding field in which I specialized—molecular genetics. One of my favorite locations in the library was near a small trolley not far from the circulation desk. There, the librarians placed newly arrived books before letting them out for circulation. It is forty-five years since those Saturday morning excursions, and I have long since forgotten almost every article and new book that I devoured with such great interest so long ago. Except for one. I remember the morning that Richard Dawkins's *The Selfish Gene* appeared on that trolley. I picked it up with interest; genes after all were my specialty. I spent that entire morning gripped by its captivating prose. Dawkins was presenting the ramifications of a godless view of the origin of our species. In his view, not only are humans and all living organisms the product of a mindless process stretching back billions of years, but our bodies are little more than "lumbering robots."

As I sat there reading through the key sections of *The Selfish Gene* on that Saturday morning, I knew that this book would attract lots of attention. Its message should be earth-shattering, I figured. As a Christian biologist, I was fairly certain that Dawkins was accurately describing the ramifications of a nontheistic view of evolutionary biology assuming, as he did, that there is *nothing* outside of material reality. In essence, Dawkins was saying that we, like all organisms, were machines created by molecules he called replicators. These replicators, made from DNA, had two functions built into their makeup. First, they contained a code that provided instructions. Second, they had a built-in capacity for making more copies of themselves. Their code provided the instructions to build the infrastructure that would enable them to thrive and be well-protected while producing multiple copies of themselves. In short, they house the instructions to build bodies. The more efficient the body they built, the more those replicators could churn out lots of copies of themselves. There was a sort of competition between different replicators and the winners were those capable of building the most effective bodies. Efficiency was everything and the losers faded from the scene like horse-drawn buggies giving way to gas-powered automobiles. With time—lots of it—some of those replicators built contraptions that were amazingly efficient: They came equipped with agile devices called legs attached to something called muscles, enabling them to move towards a source of nourishment and to flee from danger. The constructions made by certain other replicators were stationary, but what they lacked in mobility skills, they made up for in other ways: these replicators discovered how to build little solar panels called leaves so they could harness their own energy supply. Sure enough, the surface of the earth became one large mass of these solar panels, each housing their own supply of replicators mass-producing copies of themselves.[1]

As the book reached its climax, Dawkins, reflecting on the amazing success of these replicators, turned his attention to a set that produced minds that gave access to a skill enabling them to understand from whence they had come:

> They are in you and in me; they created us, body and mind; and their preservation is the ultimate rationale for our existence. They have come a long way, those replicators. Now they go by the name of genes, and we are their survival machines.[2]

We humans, he states, are robots, "lumbering robots," built to efficiently propagate our DNA molecules through eons of time. We, equipped with minds, just happen to be the fanciest and most successful machines ever produced.

As a genetics graduate student and postdoctoral researcher several years before Dawkins's book came out, I—like many others, to be sure—had reached a similar conclusion. This is where a purely materialistic view of the

1. Dawkins did not use these two specific examples, but they illustrate the point he was making.

2. Dawkins, *Selfish Gene,* 21.

biological world leaves us. It was too depressing a topic to discuss with others, though, so I seldom did, nor did anyone else I knew. But on that Saturday morning, I realized that what Dawkins had done was to write it all down in remarkably engaging prose—almost making it sound beautiful if only he hadn't been describing the engineering-like, empty-of-purpose algorithm that he said constituted the basis of human existence. Our thoughts, joys, sorrows, hopes, and dreams were all part of an illusion manufactured by a specific set of replicators as the most efficient way of propagating themselves through time. Lumbering robots, survival machines! How does one find purpose in that? I thought people would be able to see through the magic of his prose to recognize the stark reality of just how meaningless a materialistic view of reality actually is.

It had been a few years since I had waged my own internal battle with all of this, a story that I would later tell in my book *Coming to Peace with Science*. By now, however, my relationship with God through the person of Jesus Christ was the most important reality in my life, and I was sure Dawkins was wrong. But I was also glad that Dawkins had written with such engaging clarity; he had laid out the alternative to a God-centered life. Surely, I thought, this will bring about a spiritual revival; people will come to see there is no middle ground. Either Dawkins is right, and our existence is simply the product of an impersonal algorithm with no built-in meaning, or our lives are grounded in a grand eternal purpose through the God who is love. The evolutionary process was true—I was strongly convinced of that—but evolution was God's process and we have been brought into existence to live into and to thrive forever immersed in God's love. The stark choice between the two ways of thinking had never been more clear.

It will soon be five decades since Dawkins's book was written. The book is a classic, one of the best-selling biology books of all time. Although many biologists would disagree with some of the simplifying assumptions of Dawkins's "selfish gene" hypothesis,[3] in broad outline, it has stood the test of time for those who hold to a naturalistic view of reality. Indeed, almost all the human origins books written in the past twenty years or more include statements that resemble Dawkins's conclusions. For example, Harvard's E. O. Wilson, one of the most widely respected biologists of the past fifty years

3. Many would say, for example, that it is not just the gene, it is the genome. Others would say that this also is too reductionist. They would point out that genes alone can't do anything. Genes and the supportive environment (cellular and extracellular) work together to bring about the properties of the organism. Others would point out that as the embryo develops, new properties (like mind) emerge. This they would call an emergent property, one that is fundamentally different and can't be predicted from a scale-up.

writes, "Humanity, I argue, arose entirely on its own through an accumulated series of events during evolution. We are not predestined to reach any goal, nor are we answerable to any power but our own."[4]

Ernst Mayr was one of the big three architects of the evolutionary synthesis in the mid-twentieth century. Here is what he said about the existence of cosmic purpose (teleology) in the universe:

> From the Greeks on, there was a widespread belief that everything in nature and its processes has a purpose, a predetermined goal. And these processes would lead the world to ever greater perfection. Such a teleological worldview was held by many of the great philosophers. Modern science, however, has been unable to substantiate the existence of such a *cosmic teleology*. Nor have any mechanisms or laws been found that would permit the functioning of such a teleology. The conclusion of science has been that the final causes of this type do not exist. [5]

Henry Gee, the senior editor for biology (London) of the world's leading science journal, *Nature*, says evolution clearly has no plan, no memory nor any foresight. "Rather than being at the pinnacle of creation," he proclaims, "human beings are just one species on the tangled bank of Darwin's imagination." We are special in our own way, he says, but no more special than "the insects flitting above that bank" or "the worms struggling through the damp earth beneath."[6]

I could go on and describe the many statements of this sort in book after book in the human origins section of my library. Almost all contain some sort of statement about the "fact" that there is no outside nonmaterial influence affecting our evolutionary past and they imply that this is a conclusion firmly grounded in scientific data. But is this true? Does science really show that we are "unanswerable to any power but our own," as Wilson states?[7]

To the extent that Dawkins and others imply this to be true based on near-certain *scientific* conclusions alone, they are overlooking the bias that is built into the scientific process itself. Many historians of science emphasize that scientific conclusions are influenced by culture. Harvard's Steven Shapin words it like this:

4. Wilson, *Meaning of Human Existence,* 15.
5. Mayr, *One Long Argument,* 67.
6. Gee, *Accidental Species,* 12.
7. Wilson, *Meaning of Human Existence,* 15.

> I believe, of course, that there are facts . . . independent of our culture, independent of our social order, independent of our language, but I believe at the same time that when *we* make statements of fact, those statements of fact belong to our culture. So I believe completely that there is a world independent of our thoughts, but when we start to represent that world and when we talk about it, we're talking about cultural entities. What else could they be?[8]

The science of biology is embedded in its own culture. The premise within which it largely operates is that nothing exists outside of the physical world. Here is an example to illustrate my point. In a now classic survey that itself was a repetition of a much earlier survey, the religious beliefs of members of the National Academy of Sciences were surveyed. This is the most august body of leading scientists in the United States. It is an honorary, self-nominating, self-electing society, limited to 2,900 members, about 190 of whom have been awarded the Nobel Prize. Only 5 percent of the biologists believed in a personal God and immortality.[9] In Britain, the analogous organization is the Royal Society. There, of 121 biologists only three agreed or strongly agreed that a personal God existed; three were neutral, six disagreed, while the remaining 109 strongly disagreed.[10] Clearly, this is not a group of churchgoers. The prevailing culture of biological science, at least at the subconscious level, likely has an impact on these individuals' scientific conclusions, including about the purposelessness of evolution. What else could one expect?

The "godfather" of modern evolutionary genetics, I suppose, is the late Harvard professor of biology Richard Lewontin. He was a student of the great Theodosius Dobzhansky and pioneered the merging of molecular biology with evolutionary theory. In an especially forthright essay, Lewontin frankly declares that the methodology of science is a commitment to materialism. For science to progress, Lewontin maintains, the supernatural must be excluded:

> It is not that the methods and institutions of science somehow compel us to accept a material explanation of the phenomenal world, but, on the contrary, that we are forced by our *a priori* adherence to material causes to create an apparatus of investigation and a set of concepts that produce material explanations, no matter how counter-intuitive, no matter how mystifying

8. CBC Radio, "Interview with Steven Shapin."

9. Larson and Witham, "Leading scientists still reject God."

10. Stirrat and Cornwell, "eminent scientists reject the supernatural," 33.

> to the uninitiated. Moreover, that materialism is absolute, for we cannot allow a Divine Foot in the door. The eminent Kant scholar Lewis Beck used to say that anyone who could believe in God could believe in anything. To appeal to an omnipotent deity is to allow that at any moment the regularities of nature may be ruptured, that miracles may happen.[11]

This is, in effect, the way science works. In day-to-day practice, it proceeds *under the assumption* that there is nothing outside of nature—the physical world—which affects investigative outcomes. It starts with the formation of hypotheses. It then examines the predictions that emerge when the hypotheses are carefully considered. Then scientists test those predictions: are they borne out? That, in essence, is how science works. By testing and verifying predictions in this way, science has provided windows through which humankind can peer into physical reality. The fine details revealed have revolutionized society, but what they see through those windows is sometimes counterintuitive. Who could have guessed, for example, that matter is mostly empty space, or that light reaching earth would in some cases have left its source billions of years ago? Who could have guessed that DNA, the ultimate in structural simplicity, would have all the core information for making something as complex as the human brain? The knowledge gained by peering through these windows has revolutionized the world and continues to do so, offering surprises like the internet, the smart phone, and the ability to decode the DNA instructions for making the human body. Looking through one window just opens the view to another window that, in turn, soon leads to more. In the epic words of Dr. Seuss, "Oh, the places we've gone; oh, the places we'll go!"[12]

DEFINING SCIENCE

Science, as we have said, provides a window that allows us to see and understand material reality, but it works because it is grounded in a single all-important assumption. For all intents and purposes, science assumes nothing outside of the material realm exists. If ever it allows a "Divine Foot in the door" it is no longer science. At that point it is something else—but it is not science. Here is a definition from a well-respected book on the history of biology:

11. Lewontin, "Billions of Demons," 31.

12. Adapted from Seuss, *Oh, the Places You'll Go.*

> Science deals not with the gods above but with the worlds below. It does not refute the gods; it merely ignores them in its explanations of the natural world . . . Science is a way of knowing by accumulating data from observations and experiments, seeking relationships of the data with other natural phenomena, and excluding supernatural explanations and personal wishes. It has proved a powerful procedure for understanding nature.[13]

In recent years it has been especially important for legal purposes to carefully define what science is and how it works. In 1987, seventy-two Nobel laureates, seventeen state academies of science, and seven other scientific organizations submitted a brief to the United States Supreme Court on a matter related to the constitutionality of Louisiana's Balanced Treatment for Creation-Science and Evolution-Science Act. It includes what is in essence a delineation of the practical limits of science. Given the broad representation of its writers, it provides a good consensus of how science's practitioners define their discipline:

> Science seeks only naturalistic explanations for phenomena. Science is not equipped to evaluate supernatural explanations for our observations; without passing judgment on the truth or falsity of supernatural explanations science leaves their consideration to the domain of religious faith. Any body of knowledge accumulated within these guidelines is considered "scientific" and suitable for public school education; and any body of knowledge not accumulated within these guidelines is not considered scientific. Because the scope of scientific inquiry is consciously limited to the search for naturalistic principles, science remains free of religious dogma and is thus an appropriate subject for public-school instruction.[14]

SCIENTISM

The above statement is a good description of how science works on a day-to-day basis. The problem comes when scientists overstep these self-defined bounds. One example of that is demonstrated by the writing of Lewontin himself. The challenge for scientists, he stated, is to get the general public "to accept [that the] social and intellectual apparatus, Science, [is] the only begetter of truth."[15] If science, according to its top practitioners (see above)

13. Moore, *Science as a Way of Knowing*, 502–3.
14. Cited in Shermer, *Why Darwin Matters*, 96.
15. Lewontin, "Billions of Demons," 31.

is defined in a way that excludes an investigation of the supernatural, then it is inappropriate to define it as "the only begetter of truth." What if the supernatural *does* exist? How can certain scientists declare it doesn't—seemingly on scientific grounds—when, by definition, science doesn't study it? Statements of the sort we have been discussing are grounded in scientism, which the MIT physicist Ian Hutchinson, summarizes as follows: "*Scientism says, or at least implicitly assumes, that rational knowledge is scientific, and everything else that claims that status of knowledge is just superstition, irrationality, emotion, or nonsense.*"[16] Views that implicitly or explicitly are grounded in this view of the world are said to be *scientistic.*

Despite the enormous success of the scientific enterprise in revealing profound detail about the origin of the universe and the way in which it functions, it is important to emphasize that the scope of science is limited by the tools at its disposal. There is only so much the scientific method can do as it makes predictions grounded in hypotheses and then finds ways to test them. Could it be that there are values and purpose behind the universe's existence? What if there is a mind that transcends the boundaries of the universe and is in some way responsible for its laws and operation? How would a scientific analysis test for this?

MOVING BEYOND SCIENTISM

The philosopher/theologian Keith Ward has responded to the philosophical section of Richard Dawkins's book *The God Delusion.* Dawkins's fourth chapter is entitled "Why There Almost Certainly is No God"; and Ward's response is a poignant little book which uses the same title except it substitutes Dawkins's use of the word, "no" with his own word choice, "a," thereby calling his book: *Why There Is Almost Certainly a God.* Ward points out that almost all great classical philosophers through the ages have argued in favor of a statement Dawkins mockingly belittles in *The God Delusion.* Here is that statement: "There exists a superhuman, supernatural intelligence who deliberately designed and created the universe and everything in it."[17] Nothing could be more foolish, Dawkins thinks. In rebuke, Ward points out,

> Plato, Aristotle, Anselm, Aquinas, Descartes, Leibniz, Spinoza, Locke, Berkeley, Kant, Hegel . . . all argued that the ultimate reality, often hidden under the appearances of the material world of time and space, is mind or Spirit. Even the great philosophical dissenters—most notably David Hume . . . and A. J. Ayer . . .

16. Hutchinson, *Monopolizing Knowledge,* 1.

17. Dawkins, *God Delusion,* 31.

> were not materialists. Hume and Ayer thought that the ultimate realities were what they called "impressions and ideas" or "sense-data" respectively . . . Most philosophers in the world . . . have thought the ultimate reality is mind.[18]

Dawkins states consciousness (that is, mind) "comes into existence *only* as the end product of an extended process of gradual evolution . . . creative intelligences *necessarily* arrive late in the universe."[19] To this, Ward retorts: "How does he know that? What sort of evidence could there be for thinking that it is absolutely impossible for any form of consciousness to exist except the sort of consciousness that humans have?"[20]

Ward goes on to propose something he calls the God Hypothesis, which states, "There is a consciousness that does not come into being at the end of a long physical process. In fact it does not come into being at all. It did not just spontaneously appear out of nothing. It has always existed and it always will."[21] He maintains that there is no reason, grounded in science, to dispose of that which has been foundational to most of the great Western philosophers through the ages. Elaborating further on the God Hypothesis, he goes on to state:

> There is at least one mind that is prior to all matter, that is not in time and therefore is not capable of being brought into being by anything. It is the one truly self-existent reality, and the cause of all physical things. Obviously such arguments will not work if you simply assume that materialism is true. Then you will see the arguments as starting from a purely physical universe, and magically arriving at God as an extra entity just outside the universe, or snuffling around its boundary, who is made of very thin supernatural matter, and needs just as much explanation as the universe does. That seems to be exactly how Dawkins sees the arguments for God. No wonder he thinks he can dispose of them quite easily! Arguments for God do not work like that. They are arguments to show that mind is the ultimate reality, and that materialism is a delusion caused by a misuse of modern science. The arguments do not "prove" that there is one extra pseudo-physical thing in or just outside the universe. They provide good reasons for thinking that the ultimate character

18. Ward, *Almost Certainly Is a God*, 12.
19. Dawkins, *God Delusion*, 3, emphasis added.
20. Ward, *Almost Certainly Is a God*, 17.
21. Ward, *Almost Certainly Is a God*, 17.

> of the universe is mind, and that matter is the appearance or manifestation or creation of cosmic mind.[22]

Ward emphasizes that if there is such a mind, the place to start thinking about reality is not the universe and whether it requires a designing intelligence to explain its existence, but rather that the starting point should be the existence of that ultimate reality. "Arguments for God disagree with materialism at the very first step, the step of saying that matter has independent, solid, obvious reality." He goes on to say that "the reality in which we live can be best understood by postulating one kind of reality as ultimate, and as accounting for all the rich array of realities that we seem ourselves to experience."[23]

The God Hypothesis proposes that there is a Mind that knows all possible states and can distinguish between the rationally desirable (that which we call good) and rationally undesirable (that which we call bad). God acts on one fundamental principle, the principle of creating for the sake of goodness. God is the one and only cause of all existence except for the divine existence itself, which has no cause. This hypothesis is an example of the use of Occam's razor, a principle well known to scientists. It posits that in designing a hypothesis one should not multiply entities unnecessarily. As Ward notes, "God is just one entity, whereas if you have lots of elementary particles and laws, or even lots of universes, you have many entities." The simple, he states, is not more probable than the complex, but there is something esthetically pleasing in one single indivisible mind acting on one general principle (goodness).[24]

> Not only is the God Hypothesis just as probable as the materialistic hypothesis. It is more economical and elegant, and it is more comprehensive since it succeeds in accounting for purpose and value in the universe.[25]

One of the most fruitful aspects of the hypothesis, he goes on to state, is that it raises the probability (almost to the point of certainty) that the laws of nature should be conducive to the emergence of intelligent life. But what about explaining the existence of God? Here is how he responds:

22. Ward, *Almost Certainly Is a God,* 20.

23. Ward, *Almost Certainly Is a God,* 28.

24. Note that Ward's God Hypothesis is grounded in philosophy. Theology, grounded as it is in biblical revelation, especially the life, death, and resurrection of Jesus Christ, allows us to expand on the nature of God. God is not just the eternal mind that is distinct from creation, God is love and it is out of that love that God has brought all creation into existence. We will have much more to say about this in the pages that follow.

25. Ward, *Almost Certainly Is a God,* 50.

> Everything that is in time, is explained, it seems, by reference to a law and a preceding cause that brings it into being. But if we can think of something being in time, then it follows by a simple process of negation that we can think of something not being in time. Call that an eternal thing. If there is an eternal thing, then it could not be brought into being by anything else, since it could not be brought into being at all. Since it is not in time, there was never a time when it was not, and it could not possibly be brought into being.[26]

Ward shows that there are limits to how far reasoning, grounded only in scientific methodology, can take us. The assumption that there is nothing outside of physical reality is a hypothesis, but it is just that, and it is outside the realm of scientific testability.

Materialists think the idea of revelation is foolish. This is only true though if there is no outside intelligence responsible for the universe's existence. If there is a mind that seeks to communicate its goodness with those who have the capability of understanding, that changes everything. Here's what Ward has to say about this:

> If you think it is reasonable to see God as the only ultimate reality (as it is), then you will also think it is reasonable for God to reveal spiritual truth through prophets whose minds God unites closely to the divine mind. So you will be disposed to accept truth by revelation. You will not accept it blindly or without question. You must criticize, evaluate, sift and examine. But in the end it will be unreasonable to claim that God has left the world without any knowledge of God's nature and purposes, or any hope of closer knowledge of God. Only God could provide such knowledge. Revelation becomes a highly probable consequence of the existence of a supremely good God.[27]

One of the most influential scientistic writers is the blogger and population geneticist Jerry Coyne. Coyne was a student of Richard Lewontin, discussed above. To get a glimpse into his heart and that of other materialists like him, I highly recommend reading his moving tribute to Lewontin written on the day after Lewontin passed away in July 2021.[28] Coyne has a keen sense of goodness, even though he thinks it emerges from within a blind, purposeless set of laws that govern the universe. It certainly, Coyne thinks, does not emerge through revelation:

26. Ward, *Almost Certainly Is a God*, 52.

27. Ward, *Almost Certainly Is a God*, 128.

28. https://whyevolutionistrue.com/2021/07/05/dick-lewontin-1929-2021/.

> Truth is simply what is: what exists in reality and can be verified by rational and independent observers. It is true that DNA is a double helix, that the continents move, and that the Earth revolves around the Sun. It is not true, at least in the dictionary sense, that somebody had a revelation from God. The scientific claims can be corroborated by anyone with the right tools, while a revelation, though perhaps reflecting someone's real perception, says nothing about reality, for unless that revelation has empirical content, it cannot be corroborated.[29]

This is grounded, of course in Coyne's belief that there is nothing outside of material reality and that all truth must be scientifically verified. Revelation, he thinks, cannot be "verified by rational and independent observers" and thereby cannot be considered a source of truth. Coyne considers science and religion as outright competitors "in the business of finding out what is true about our universe." It is, he says, a "battle in the war between rationality and irrationality."[30] Coyne is scientistic, but he is also an exceptionally good scientist. Interestingly, he describes the scientific process a little more broadly than those two definitions with which I began this discussion. He points out that even historical analysis can count as science since it reveals truths about the natural world:

> And sometimes the "tests" of hypotheses don't involve experiments, but rather observations—often of things that occurred long ago. It's hard to do experiments about cosmology, but we're completely confident in the existence of the Big Bang because we observe things predicted by it, like the expanding universe and the background radiation that is the echo of that event. Historical reconstruction is a perfectly valid way of doing science, so long as we can use observations to test our ideas (this, by the way, makes archaeology and history disciplines that are, in principle, scientific).[31]

So, according to Coyne, we can learn things about the natural world through historical analysis. For example, this approach allows investigators to become confident about certain events like the big bang event 13.8 billion years ago. But Coyne, justifiably in my opinion, extends scientific study to include work outside of that which we normally call the natural sciences. He includes, for example, the analysis and conclusions archaeologists make regarding ancient civilizations, that linguists make about historical

29. Coyne, *Faith versus Fact*, 29.

30. Coyne, *Faith versus Fact*, 1.

31. Coyne, *Faith versus Fact*, 32.

relationships between languages, and that art historians make when they determine that a particular painting is a forgery.[32] Science, Coyne tells us, is a tried and trustworthy method of determining truths about the natural world and even historical analysis can legitimately be called a form of science. But Coyne, sticking to the traditional view, and that of his mentor, Lewontin, most certainly would limit his definition to the study of that which exists in the material realm and indeed explicitly states this in his book *Faith versus Fact.*[33]

RESURRECTION AS A HISTORICAL EVENT AND AN ALTERNATIVE HYPOTHESIS ABOUT REALITY

What if historical analysis is used to explore questions that relate to the possibility of a nonmaterial reality? We have established that by the standard definitions for science, this would not be called science. Fine. But if historical analysis is a legitimate way of exploring reality, as Coyne indicates, might not its tools be valid for testing a hypothesis about the actual occurrence of a specific historical event that has a nonmaterial (supernatural) basis? Highly sophisticated investigators at first-rate universities have been using historical reconstruction tools to test such hypotheses. Consider, for example, the resurrection of Jesus. N. T. Wright is senior research fellow at the University of Oxford. In essence, his book *The Resurrection of the Son of God* is an 850-page analysis of the historical hypothesis that the resurrection actually took place. As he carefully lays out the data and considers the options on how they can best be understood, he argues that only one hypothesis stands out as being reasonable, namely that "the tomb really was empty and that people really did meet Jesus alive again."[34] The importance of this, if true, cannot be overstated. The existence of a loving, nonmaterial reality would take us out of the scientistic realm and usher us into an existence that extends beyond the universe.

Here is a summary of the data that led Wright to the above conclusion. First, given the rapidity with which the notion of resurrection appears to have been accepted by a sizeable number of people, it was important to consider whether something like this would have already been on the "cultural radar screen" at the time. If resurrection stories were somewhat common or were expected to happen, it would not be difficult to imagine made-up or imagined stories surfacing quite quickly following the death of a cult

32. Coyne, *Faith versus Fact,* 39.

33. Coyne, *Faith versus Fact,* 64.

34. Wright, *Resurrection of the Son of God,* loc. 381.

figure. With that in mind, Wright carefully explores the Greek, Roman, and Egyptian literature of the era and demonstrates any views of an afterlife did not include bodily death followed by restored life in a transformed body. In essence, he demonstrates, the notion of the resurrection of a body—once dead—coming to life again in some altered physical form did not exist in their belief system or their literature. Within the Jewish community, the Pharisees *did* believe that death was not permanent. They thought that death led to an intermediate state which would at some future time be followed by a restored body. They did not think, however, that this final restorative event had yet happened to anyone, and although they thought their bodies would eventually come to life again, they were not expecting anything like the mystical body of the resurrected Jesus. According to Wright, the notion of a sometimes-difficult-to-recognize resurrected person "who comes and goes through locked doors, who is solidly physical, with wounds still visible and yet seems to belong in two dimensions at once"[35] was an as yet unimagined concept—not only to Jews but to the surrounding cultures as well. Furthermore, although Jews were expecting a messiah, they were not expecting he would die an inhumane death followed by a miraculous resurrection into an altered body-state. Yet this is what those first Christians believed happened—a humiliating death quickly followed by not a resuscitation but a resurrection of an altered, seemingly mystical body which was, nonetheless, clearly Jesus'.

Second, if the resurrection story was made up, there would be considerable effort on the part of the Gospel writers to show how it fulfilled scriptural prophecy. After all, Wright points out, all four Gospels emphasize different ways in which Jesus' life and death fulfilled specific Scriptures. Yet when it comes to the resurrection, a story "crying out" for verification, the Gospels turn almost silent with regard to fulfillment of specific scriptural predictions, although certainly Scriptures existed that might be interpreted that way retroactively. The Gospel of John notes that when two of Jesus' disciples ran to the tomb, "They still did not understand from Scripture that Jesus had to rise again from the dead."[36] It would have been so easy for the writer to have slipped in a specific Scripture to justify his point; instead, John is silent. As the Gospel of Mark tells the story of Jesus' death in chapters twelve through fifteen it is full of references to the way in which events fulfill Old Testament passages. But moving into an account of the resurrection, Mark's referencing of scriptural predictions turns silent. Not once in Matthew's narrative about the resurrection, Wright notes, do we see Matthew

35. Wright, *Resurrection of the Son of God*, loc. 12241.

36. John 20:9.

using the familiar line scattered throughout his account: "All this took place so that it might be fulfilled what was spoken by the prophets . . ." And the master storyteller, Luke, provides a wonderful narrative of the conversation between two unnamed disciples and the resurrected Jesus on the Emmaus road. Luke tells the reader that the primary focus on that journey was how the Scriptures had predicted what had happened. This would have been the perfect opportunity for Luke to specifically relate *how* Scriptures had predicted resurrection, but he doesn't. Similarly, Luke writes that the angels at the tomb declared that the death and resurrection (after three days) "must" happen. Then, as he discusses the upper room experience, Luke emphasizes that Jesus expounded on how to understand the Scriptures that pointed to his death and resurrection after three days. The telling of any of these stories would have represented the ideal opportunity for Luke to "sell" this to his readers if he was making up a story or passing along a legend that he wanted his readers to accept as truth. It is not that such Hebrew Scripture-based references did not exist in the Christian community, and Wright examines those specifics extensively in his book, but the point is that when the Easter narrative is told at length in each of the four Gospels, there is simply no attempt to make the resurrection more believable.

Third, if the resurrection story was made up, one would expect that it would be told in a manner that reassures the Christian in a style centered on hope. That, after all, is how it is often told at Easter today: because Jesus is risen, we are justified in the hope that we too shall live eternally. Indeed, other New Testament writers, especially Paul, put much emphasis on this aspect of the meaning of the resurrection story. However, as Wright puts it: "If Matthew, Mark, Luke and John wanted to tell stories whose import was 'Jesus is risen, therefore you will too,' they have done a remarkably bad job of it."[37] Instead, the Gospels' stories are told in the context of a commission in the present world: "Jesus is risen, therefore you have work to do." So, the resurrection story is not told in a manner that seeks to convince, despite its radical departure from expectations in both the pagan and Jewish world. Rather it just assumes everyone knows it is true, and it focuses on the task at hand.

Fourth, if the resurrection story was made up, we would expect that the resurrected Jesus might be exalted to a position of divinity or some other sort of heavenly glory. Surely, Wright suggests, the story would be more believable that way. Instead, the resurrected Jesus appears as a human among humans, but with a transformed body that makes it difficult to recognize who he is. Indeed, with this in mind Matthew frankly tells us that

37. Wright, *Resurrection of the Son of God*, loc. 12110.

"some doubted" even as they were in the presence of the risen Jesus. Luke articulates that two of his disciples walked along with him for several miles, without recognizing him until the moment he broke the bread as they sat down to eat. And then there is the Gospel of John. Here we are told that the disciples did not know it was Jesus calling out to them giving instructions on where to cast their net. Put simply, these stories, filled as they are with doubt about whether this person was really Jesus, are not the sort of stories that one would predict would be told if they were made-up stories designed to convince others that the unlikely had really happened.

Fifth, for a made-up story so hard to accept, one would expect considerable effort would have been made to ensure it was reported by impressively credible witnesses. Instead, all four of the Gospels tell us that the first witnesses to the empty tomb and the first to sight the resurrected Jesus were women. Wright notes, "Even if we suppose that Mark, [for example], made up most of his material . . . it will not do to have . . . *women be the ones who find it* . . . women were simply not acceptable as legal witnesses. We may regret it, but this is how the Jewish world (and most others) worked."[38]

There is another possible objection to the hypothesis that needs to be examined: an honest misunderstanding. What if, for example, the tomb was empty either because of theft or confusion about where the body had been laid? Alternatively, it is not unreasonable to suspect that out of the depth of their despair that certain of Jesus' followers had visions of a restored, resurrected Jesus. In theory, either of these could lead to sincere reports that Jesus had risen from the dead. However, an empty tomb alone, although necessary for the believability of the resurrection story, would not have been sufficient for it to have taken hold in a larger community; quite simply it would have been too likely that someone had just come and taken the body. Similarly, sightings of the resurrected Jesus would have been necessary for the emergence of a belief in the resurrected Lord, but likely not sufficient for it to be accepted—not if the body was still in the tomb. False dreamlike visions by perfectly sincere people do occur on occasion, so this in itself, although necessary, would not be sufficient to convince a sizeable contingent that the resurrection had really taken place. In order for the resurrection story, sincerely told, to be sufficient to lead to wide acceptance, both an empty tomb and sightings of the resurrected Jesus would have been necessary. Since detailed descriptions of both sets of events (an empty tomb, with neatly folded burial clothes and many sightings) are described as factual accounts, it is unlikely that it emerged out of a sincere and honest misunderstanding.

38. Wright, *Resurrection of the Son of God*, loc. 12206.

There is a strong consensus that the earliest writings that we have referring to the resurrection are those of Paul. Likely, most scholars agree, based on strong historical evidence, Paul was writing within about twenty-five years of the death of Jesus. Paul testified to the reality of the resurrection account and stressed that five hundred persons—many of whom were still alive and available to serve as witnesses—could confirm its authenticity. More importantly though, Paul based his whole elaborate theology on the resurrection of Jesus. This is laid out in a remarkably articulate manner in his First Letter to the Corinthians, with chapter 15 as its centerpiece. Wright analyzes in detail how the letter gradually works its way to this zenith, and then dissects the fifteenth chapter—section by section—showing just how central the factuality of the bodily resurrection is to purpose and meaning in life. It is true that many biblical scholars and theologians of the twentieth century proposed that Paul was writing metaphorically about the resurrection, but Wright carefully lays out each of the statements that Paul makes about it, and illustrates in step-by-step fashion just why this metaphorical interpretation is unlikely. Furthermore, it is important to stress that a careful reading of that fifteenth chapter (indeed, the whole book) makes it seem highly unlikely that Paul (or someone writing in Paul's name decades later) could have made all of this up—it seems too full of sincerity and too accountable to others for its veracity to have been vacuous.

Despite how it may seem, my purpose in the above is not to convince you, the reader, that the resurrection really did occur. This is not a book of Christian apologetics. Rather, my chief motivation is to emphasize that the hypothesis—that Jesus died and rose again—holds up well to science-like scrutiny. Not only has it *not* been falsified, there are logical reasons grounded in historical analysis for accepting it as a viable hypothesis, indeed even as the best explanation of the data at hand.

THE WORSHIP OF JESUS AND AN ALTERNATIVE HYPOTHESIS ABOUT REALITY

Wright's analysis is just one example of the sort of study that sheds light on this hypothesis that the resurrection really did occur. The work of the recently deceased head of the Centre for the Study of Christian Origins at the University of Edinburgh, Larry Hurtado, provides a second example that I will discuss briefly here. In another seven-hundred-page tour de force, consisting of more than 1,200 references and based on twenty-five years of research at the Universities of Manitoba and Edinburgh, Hurtado examined the origin of Christianity as it developed in its first 150 years.[39] He explores the question of how and when it came about that early Christian devotion to Jesus—as God—and thereby worthy of worship, came about. It was remarkably early. Like N. T. Wright in his study, Hurtado's work is exhaustive, and he shows how the evidence strongly points to this unique form of devotion to Jesus as almost having been universally in place within Christian communities by the time Paul started writing his letters within a little more than two decades after the death of Jesus. It is difficult for us laypersons—examining that culture from outside its time and space—to fully appreciate how radical the notion of worshipping this person, Jesus, as God, would have been. The Jewish people were strict monotheists. They had laws in place, dating back hundreds of years, one of which was, "Thou shalt not have any gods before me." They were expecting a messiah, but there was no indication that person would be God himself and thereby worthy of worship. Nonetheless, in a manner that is completely analogous to how *scientific* analysis proceeds, Hurtado examines various possible causes for this aboutface. He demonstrates that the most likely basis of the remarkably explosive nature of this development—thinking of Jesus as God—would have been "revelatory religious experiences." These would have included a belief in the veracity of the resurrection. Unlike Wright's work, Hurtado does not analyze the likelihood of the resurrection itself. Instead, his focus is on *the belief* that it occurred as well as a belief in other revelatory experiences. That belief, he shows, was the most likely cause of the almost instantaneous devotion that early Christians had to the notion of Jesus as God. That belief, his data indicate, was uniformly widespread in the community within a few decades.

RELIABILITY OF THE GOSPEL ACCOUNTS AND AN ALTERNATIVE HYPOTHESIS ABOUT REALITY

The third example of a major historical analysis related to the origin of belief in the resurrection is the work of Richard Bauckham, in which he analyzed the question of whether the Gospels are based on eyewitness testimony. Richard Dawkins insists in *The God Delusion* that the New Testament Gospels are not reliable accounts of what actually happened in the life of Jesus, and he states this has been the consensus of "scholarly theologians" since the nineteenth century:

> Ever since the 19th century, scholarly theologians have made an overwhelming case that the gospels are not reliable accounts of

39. Hurtado, *Lord Jesus Christ.*

> what happened in the history of the real world. All were written long after the death of Jesus, and also after the epistles of Paul, which mention almost none of the alleged facts of Jesus' life. All were then copied and recopied through many different "Chinese Whispers generations" by fallible scribes who in any case had their own religious agendas.[40]

Dawkins is likely correct that scholarly theologians of the nineteenth century thought that the Gospels are not reliable accounts of real events. However, he is not correct that this view is the consensus of biblical scholars today. The alternative view is lucidly laid out by Bauckham's analysis:

> It is the contention of this book that, in the period up to the writing of the Gospels, gospel traditions were connected with named and known eyewitnesses, people who had heard the teaching of Jesus from his lips and committed it to memory, people who had witnessed the events of his ministry, death, and resurrection, and themselves had formulated the stories about these events that they told.[41]

His book has been positively reviewed; indeed, it was the subject of a whole issue of the *Journal of the Study of the Historical Jesus*, where the reasonableness of its conclusions was largely accepted.[42] Bauckham's work is broad and far-reaching. For example, he analyzes a database of the comparative frequency of the given names of about three thousand persons from the period. Not surprisingly, some names (like Joseph and Simon, for example) are very common. Other names (like Sabbataus or Pappus) are quite rare. Palestinian Jews, in particular, have a distinctive frequency distribution and it is significantly different than that of Jews from outside of Palestine. Here is what is fascinating about that observation: Bauckham shows that "the names of Palestinian Jews in the Gospels and Acts coincide very closely with the names of the general population of Jewish Palestine in this period, but not to the Jews of the Diaspora."[43] If the stories had been made up much later by persons outside of Palestine, we would not expect that New Testament characters, with made-up names, would be found with a frequency-manner that corresponded so closely to that of Jews in Palestine. This implies the Palestinians in those stories were real people.

40. Dawkins, *God Delusion*, 92.

41. Bauckham, *Jesus and the Eyewitnesses*, 93.

42. *Journal for the Study of the Historical Jesus*, vol. 6, #2, 2008.

43. Bauckham, *Jesus and the Eyewitnesses*, 73.

There is another even more significant observation, however. Consider the following: If the Gospel stories were largely made up, one could accept that as names were chosen for individuals in the fictional account, the common names chosen would largely mirror the names in use at the time. What would not be expected however for a set of concocted stories is that the distribution of rare names in fictional stories would so closely correspond (in frequency and pattern) to that in the real world. It seems highly contorted to imagine that fictional storytellers would be able to match up their pattern and frequency of rare names so well, Bauckham asserts. More likely, he thinks, is that the reason the frequency matches up so closely to the real world is that the stories were based on the accounts of real people and actual incidents.

Bauckham goes on to show that the standard practice of historical writers of this region was to ensure accounts were grounded in authentic eyewitness recollections and not (if possible) secondhand information. By examining other similar, but secular, non-biblical writings, he illustrates that "the climate of the day" required that historical accounts not be based on hearsay or renditions from a community not closely tied to eyewitnesses. Although anecdotal, here are two examples of statements that have implications for how the gospel accounts were assembled. The first is from a Greek author, Papias, who lived from 60 to 130 AD. Papias wrote five books near the turn of the century that were entitled *Expositions of the Sayings of the Lord.* Although all copies of these manuscripts are now lost, brief portions exist by having been incorporated into the writing of others while the texts still existed. Papias described the way he obtained his material as follows:

> I shall not hesitate also to put into properly ordered form for you (sing.) everything I learned carefully in the past from the elders and noted down well, for the truth of which I vouch. For unlike most people I did not enjoy those who have a great deal to say, but those who teach the truth. Nor did I enjoy those who recall someone else's commandments, but those who remember the commandments given by the Lord to the faith and proceeding from the truth itself. By chance anyone who had been in attendance on the elders should come my way, I inquired about the words of the elders—[that is,] what [according to the elders] Andrew or Peter said (*eipen*), or Philip, or Thomas or James, or John or Matthew or any other of the Lord's disciples, and whatever Aristion and the elder John, the Lord's disciples, were saying (*legousin*). For I did not think that information from

> books would profit me as much as information from a living and surviving voice (Eusebius, Hist. Eccl. 3.39.3-4).[44]

Although Papias was not an author of one of the primary Gospels, his life overlapped with some of the last remaining apostles and his writing indicates just how important it was for him to rely on eyewitness accounts whenever possible. It is reasonable to conclude this would have also been true of the writers of the canonical Gospels. Indeed, in at least one case (the Gospel of John), the writing was likely completed only a decade or so earlier. Papias actually interacted with those who knew Jesus, and many scholars believe the person he refers to as "the elder John" was the actual author of John's Gospel.

Papias was a contemporary of Polycarp (69–155 AD), another early church father who had direct contact with some of the apostles. Writing a few decades after Polycarp had died, Irenaeus (130–202 AD) wrote fondly of his memories of Polycarp as he told the stories of original apostles he knew. I love this reflection of Irenaeus written in a letter about 190 AD:

> When I was still a boy I saw you [the person to whom this letter is written] in lower Asia in the company of Polycarp, faring brilliantly in the imperial court and trying to secure his favour. For I distinctly recall the events of that time better than those of recent years (for what we learn in childhood keeps pace with the growing mind and becomes part of it), so that I can tell the very place where the blessed Polycarp used to sit as he discoursed, his goings out and his comings in, the character of his life, his bodily appearance, the discourses he would address to the multitude, how he would tell of his conversations with John and with the others who had seen the Lord, how he would relate their words from memory; and what the things were which he had heard from them concerning the Lord, his mighty works and his teaching, Polycarp, as having received them from the eyewitnesses (*autoptōn*) of the life of the Logos, would declare altogether in accordance with the scriptures. To these things I used to listen diligently even then, by the mercy of God which was upon me, noting them down not on papyrus but in my heart.[45]

Bauckham cites a series of other examples that illustrate that there was much concern about the importance of accuracy during the era in which the Gospels were written. If the accounts were not based on actual eyewitness

44. Cited in Bauckham, *Jesus and the Eyewitnesses*, 293.

45. Cited in Bauckham, *Jesus and the Eyewitnesses*, 295.

reports, the perception is that they should at least have come from individuals who had directly interacted with those eyewitnesses. Such accounts, subject to this level of scrutiny, would have included the resurrection of Jesus.

Too many scientists dismiss the possibility of Jesus' resurrection out of hand without carefully considering the sort of analysis that has been done to convince many rigor-demanding scholars of history that it really happened. Too many scientists think that because the resurrection was supposedly a one-time, non-repeatable event outside of what can be studied by science, it must not be true. But the origin of the universe is not repeatable either. The big bang is the best explanation of many predictions that are borne out following detailed analysis. Singularities of this sort require an expansive analysis of indirect considerations. The resurrection is like this, except it has one extra constraint: Given the definition that scientists ascribe to their discipline (science only studies actions that have material causes), investigation of a nonmaterial basis for the resurrection cannot be considered scientific. That is fine—science should be defined *as it is actually practiced.* However, that does not make the resurrection beyond the realm of investigation, and even more importantly it certainly does not make it untrue. But if the resurrection really did take place, it changes everything. Jerry Coyne, in effect, acknowledges this, calling it the "linchpin of virtually all Christian faith." To make his point he quotes Paul:

> Now if Christ be preached that he rose from the dead, how say some among you that there is no resurrection of the dead? But if there be no resurrection of the dead, then is Christ not risen: and if Christ be not risen, then *is* our preaching vain, and your faith *is* also vain.[46]

To build his case for unbelief, Coyne writes that Christianity must be grounded in real historical events. He is right, but what he does not do is to carefully consider the scholarship that addresses the question of whether it was a real event. If it was, it would be the linchpin pointing to a nonmaterial reality—the mind to which Keith Ward refers. Indeed, if true, it would explain why, in an astonishingly short span of time after Jesus' death, a binitarian[47] devotion to Jesus as God developed quite uniformly throughout the region. Not only that, but it would also explain the basis of the similarities and differences in the gospel story: the stories emerged from the various accounts memorialized by different groups of eyewitnesses. If the resurrection really happened, and there are extremely good "science-like" reasons

46. Coyne, *Faith versus Fact*, 45. Coyne quotes from 1 Corinthians 15:12–14, KJV.

47. The term "binitarian" refers to worship of Jesus along with the worship of God the Father.

to think it did, then science, because it only studies naturalistic principles, provides only a partial access to reality.

CROSSING THE THRESHOLD

The late Southern Methodist University theologian William Abraham compared entry into that more complete reality to crossing over a front-door threshold and entering a house for the first time:

> From the outside there is much to be seen externally; but far more remains hidden from sight. Step inside across the threshold and a whole network of rooms, staircases, closet, and kitchen is available. One has crossed into another world. . . . One may or may not step across the threshold; however, the step across the threshold is singular and absolute; and, once across, there is a new awareness of phenomena previously hidden from view.[48]

Even though we can ground revelation in a rationale heavily embedded in the historical reasonableness of the life, death, and resurrection of Jesus Christ, divine revelation, Christians believe, is by no means limited to events in ancient history. Once one accepts the reasonableness of these events, divine revelation—an illumination and enlightenment of the Holy Spirit—follows repentance and a commitment to virtue. As Abraham put it, spiritual exercises help to "focus the attention and purify the doors of perception," enabling us to see God.[49]

> What is at issue is purity of heart rather than ingenuity in framing hypotheses. The former is open to all given to certain assumptions about creation and prevenient grace. The latter is restricted to those who have the relevant ingenuity and creativity. If commitment to divine revelation is pivotal for a robust spiritual life, then it would be odd indeed if God left us to such limited expertise.[50]

All that sounds crazy to Dawkins, Gee, Wilson, Lewontin, and Coyne. Stephen Pinker, in his book *Enlightenment Now*, expresses the same sentiment:

48. Abraham, *Crossing the Threshold*, 86.
49. Abraham, *Crossing the Threshold*, 76.
50. Abraham, *Crossing the Threshold*, 77.

> To take something on faith means to believe it without good reason, so by definition a faith in the existence of supernatural entities clashes with reason.[51]

They all think Christians believe what they do for reasons that resemble why a child believes in Santa Claus. Pinker goes as far as to state that belief in the unreasonable is the very definition of faith. He is wrong. This is simply not the way Christianity works. There is an intellectual basis for faith, and we have just examined one aspect of the sort of data that informs faith and makes it reasonable. There are many more.[52]

The experiential dimension of faith is further described by John Cottingham, professor emeritus of philosophy at the University of Reading in Britain. He stresses that although intellectual assent is part of the basis of conversion into a life of Christian faith, it is not the only factor. In the gospels, he points out, the change of mind *(metanoia)* is more than just a rational decision—it is a radical reorientation. In essence, the Gospels start off with the call to repentance, first by John the Baptist and then by Jesus. As the Gospel of John proceeds we see, in the dialog of Jesus with Nicodemus, the striking image of the need for a complete rebirth.[53] Cottingham states that "God always allows space for free human choice in rejecting or accepting his call."[54] Cottingham goes on to quote from Blaise Pascal:

> Our religion asserts . . . that God . . . has hidden himself from direct human knowledge; indeed, the very name he gives himself in Scripture is *Deus absconditus*—the hidden God (Isaiah 45:15) . . . God has appointed visible signs . . . to make himself known to those who seek him sincerely; and . . . he has nonetheless veiled those signs in such a way that he will be discerned only by those who seek him with all their heart.[55]

Elaborating on Pascal's words, Cottingham adds:

> Conversion, if [Pascal] is right, is not a coercive process engineered by demonstrations of power, but is a response of the whole person—intellectual, emotional, moral, spiritual—that enables what was hitherto hidden to come to light. The process

51. Pinker, *Enlightenment Now*, 30.

52. There are many books that lay out the reasonableness of the Christian faith. Two examples geared towards a general audience include Keller, *Reason for God* and McLaughlin, *Confronting Christianity*.

53. John 3:1–21.

54. Cottingham, *Why Believe?*, loc. 1632.

55. Pascal, *Pensees*, cited in Cottingham, *Why Believe?*, loc. 1634.

> is not one of being brought up short by new scientific evidence or paranormal events, but the working of an interior change that generates a new openness. Nothing can force acceptance unless we have "ears to hear." And what is heard is not a barrage of confirmatory data, but a message that needs to be understood. It is, as the second epistle of Peter beautifully puts it, a *word*—one that must be "heeded, as a light that shines in a dark place, until the day dawns and the morning star rises in your hearts."[56]

ACADEMIC RIGOR AND SCIENTISM?

With this in mind, it is important to consider whether there is more than academic intellect involved in the choice of scientistic spokespersons not to consider that which "shines as a light in a dark place." The rapid dismissal of the intellectual basis of Christianity without carefully considering whether it may have reasonable basis seems so uncharacteristic of the detailed logic of these great thinkers when their thinking is restricted to scientific matters. The fact is that when it comes to the dismissal of Christian theism, these scientists demonstrate little understanding of that which they are critiquing. None of them shows any evidence, for example, of being familiar with any of the three outstanding biblical scholars (Wright, Hurtado, and Bauckham) just discussed. There is no sign that they have read some of the really fine Christian apologetic analyses—books like *The Reason for God* by Timothy Keller or *Confronting Christianity* by Rebecca McLaughlin. If they were really seeking to relay an understanding of the topics under discussion in their books, they could have referred to *Why Believe?*, by John Cottingham. They don't cite any of the other outstanding biblical scholars who describe how the Old Testament Scripture functions. Scholars like Denis Lamoureux, John Walton, Tremper Longman, Scot McKnight, and Richard Middleton, whose work has been so important to many of us, are simply ignored. They also largely ignore our best theologians—the thinking of great theological minds like Alister McGrath, Jürgen Moltmann, and Dietrich Bonhoeffer is left unexplored. And although one of them (Coyne) mentions John Polkinghorne, he does it in a manner that does not engage any of his theological work—his only point is that Polkinghorne thinks that Christianity must be grounded in truth (which Coyne defines as empirical reality). Shockingly (to me), in his review of Coyne's book, Pinker expresses appreciation for Coyne having taken the time to become educated "in the works of the most 'sophisticated' theologians and pays them the respect of evaluating their

56. Cottingham, *Why Believe?*, loc.1675.

arguments by the ordinary standards of intellectual discourse."[57] But he did not do this. At best, he did a flyover—missing almost everything—and for Pinker to think otherwise is a clear indication that he, the author of his own six-hundred-page anti-theistic tome on humanism, has not even begun to do that.[58] In short, these scholars dismiss Christianity as baseless with little more than a perfunctory knowledge of what it is about. Richard Dawkins's *The God Delusion* likely illustrates this best. *The New York Times* review points to the chapter where Dawkins states that he has come to "the central argument in my book." What is that argument? "At heart," the review states, "this argument is an elaboration of the child's question 'But Mommy, who made God?'"[59] He states that since the hypothetical creator made a very complex universe, then obviously that creator would need to be even more complex than the vast amount of complexity in the universe. Dawkins, thinking he has come across a quasi-mathematical method to show the infinitesimally low probability of the existence of God, is mystified that the argument has not been well-received by theologians. Perhaps, if the "eyes of his heart" were opened more fully,[60] and the veil of which Pascal spoke pulled back, Dawkins might begin to see reality in a whole new light. If he did, he'd realize that there are great minds who have taken the matter far beyond the delusion of his childhood view of God. However, even at that, *secular* critics were disappointed by Dawkins's inadequate scholarship of whether the existence of God is really a delusion. *The New York Times* review went on to say:

> If God is indeed more complex and improbable than his creation, does that rule him out as a valid explanation for the universe? The beauty of Darwinian evolution, as Dawkins never tires of observing, is that it shows how the simple can give rise to the complex. But not all scientific explanation follows this model. In physics, for example, the law of entropy implies that, for the universe as a whole, order always gives way to disorder; thus, if you want to explain the present state of the universe in terms of the past, you are pretty much stuck with explaining the probable (messy) in terms of the improbable (neat). It is far from clear which explanatory model makes sense for the deepest question, the one that, Dawkins complains, his theologian friends keep harping on: why does the universe exist at all?

57. Pinker, "Untenability of Faitheism."

58. Pinker, *Enlightenment Now.*

59. Holt, *Beyond Belief.*

60. Eph 1:18.

> Darwinian processes can take you from simple to complex, but they can't take you from Nothing to Something. If there is an ultimate explanation for our contingent and perishable world, it would seemingly have to appeal to something that is both necessary and imperishable, which one might label "God."[61]

And even more succinctly, the former PhD student of Jerry Coyne, H. Allen Orr, wrote:

> It's one thing to think carefully about religion and conclude it's dubious. It's another to string together anecdotes and exercises in bad philosophy and conclude that one has resolved subtle problems. I wasn't disappointed in *The God Delusion* because I was shocked by Dawkins's atheism. I was disappointed because it wasn't very good.[62]

The fact is that Dawkins has a fine scientific mind, and when he's writing only about biology his work constitutes respectable scientific integrity and logic. In *The God Delusion,* however (and indeed in all his books where he addresses the possibility of a nonmaterial reality), his level of rigor evaporates. More importantly however, given the statements about what constitutes science discussed above, he does the very thing that the seventy-two Nobelists writing to the United States Supreme Court claimed science does not do.[63] Science, he tells his host of readers *is* "equipped to evaluate supernatural explanations."[64]

It is almost fifty years since I first saw the sparkling copy of *The Selfish Gene* on the new-releases trolley in the university medical library. The impact of that book has been nothing short of astounding. Orr, the evolutionary geneticist who was so critical of *The God Delusion,* declared, in complete contrast, that *The Selfish Gene* was "the best work of popular science ever written."[65] But was it, at heart, a science book? Had that book restricted itself to the findings of the science of biology, or was it undergirded by an unexamined and untested philosophy that Dawkins simply assumed to be true? Is there another story that needs to be told? Is there a story that depicts

61. Holt, "Beyond Belief."

62. Orr, "God Delusion."

63. "Science is not equipped to evaluate supernatural explanations for our observations; without passing judgment on the truth or falsity of supernatural explanations science leaves their consideration to the domain of religious faith." Cited in Shermer, *Why Darwin Matters,* 96.

64. The quotation is from the letter to the Supreme Court discussed earlier in this chapter.

65. Orr, "Mission to Convert."

the origin of our species in scientifically accurate terms, but also does justice to the beauty and reasonableness of the Christian faith? That is the question to which we now turn.

2

The Origin of Our Genus, *Homo*

A COOPERATIVE MINDSET

One of the most sentimentally engaging videos I have ever seen is that of a dying fifty-nine-year-old chimpanzee called Mama.[1] Lying in the fetal position, looking sick and about to die, Mama turns down all food and drink until eighty-year-old primatologist Jan van Hoof enters the cage. Not having seen Mama in a long while despite their forty-year-long relationship, he calls out to her. Slowly her eyes open as she begins to recognize the voice. Then suddenly, realizing who he is, she breaks into a massive full-lipped grin as she weakly stretches out her hand to lovingly groom his hair and gently pat his head one last time. Except for her chimpanzee vocalizations, the touch of her fingers, the glow of her facial expressions, and the one last hug make the interaction not unlike that which occurs between two humans in hospitals around the world every day. Chimpanzees are very much like us.

It is not just that we share the same set of emotions,[2] chimpanzees outshine us in certain aspects of memory and intelligence. Another video shows a chimpanzee at his primate research home in Japan, where he has learned a computer game carrying out feats of number recognition and short-term memory that no human would be able to duplicate.[3] Chimpanzees have the capability of planning ahead, picking up tools that they will use much later for digging into a termite mound or to acquire honey from a beehive.[4] In

1. https://www.youtube.com/watch?v=INa-oOAexno.
2. See de Waal, *Mama's Last Hug*.
3. https://www.youtube.com/watch?v=zsXP8qeFF6A.
4. See de Waal, *Mama's Last Hug*, 135.

one case of forethought, a bonobo was seen bearing a heavy rock on her shoulders with her infant holding on just behind, as she journeys to a place where she knows there are nuts that can be cracked open using the rock.[5] In one lab study, chimpanzees, orangutans, and two-and-a-half-year-old children were tested for spatial, quantitative, and causational skills. Remarkably, the great apes and human children were about equal for almost all the skills.

There is one exception, though, and it is not only noteworthy, it is likely the key to understanding what it is that makes humans profoundly different from the great apes (see Figure 2.1). When it comes to the sort of skills required for functioning together in a cooperative manner, the children were off-the-scale superior to their great ape cousin species. From the earliest stages of their lives, humans are uniquely "wired" for cooperation.

FIGURE 2.1

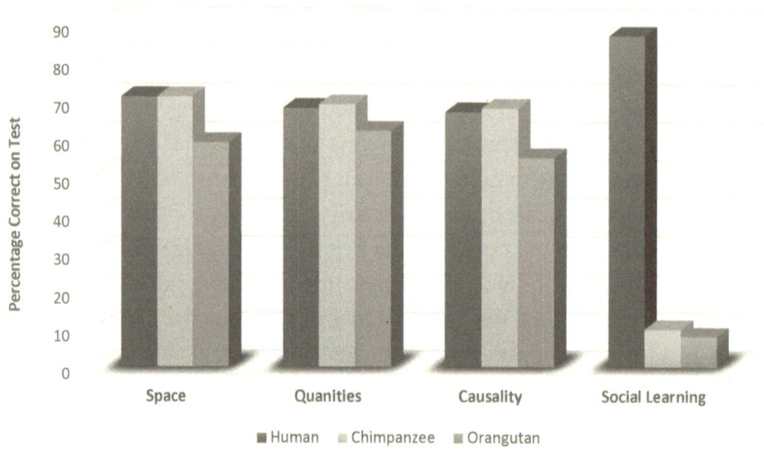

Average performance on four sets of cognitive tests with chimpanzees, orangutans, and toddlers.[6]

The fact that sociality begins to manifest itself so early in human development illustrates that we are genetically and neurologically wired to live in a manner that is in tune with the goals and ambitions of others. Duke University psychologist Michael Tomasello shows that children as young as fourteen to eighteen months form pairwise partnerships that enable them to work effectively towards a common goal. Indeed, so close is the paired

5. De Waal, *Are We Smart Enough?*, 214.

6. Redrawn with permission, from Henrich, *Secret of Our Success*, 15.

relationship that forms in children that he calls it a "we" that functions in essence as a unit. Each member of the "we" knowingly acts *with* the other as they move together towards their single purpose, while being fully aware that the other is equally aware of that single unified goal. But there is also an "I" function, in which each individual has her own role to play and her own perspective, even as each is entirely cognizant of the other's different but complementary role.[7] This phenomenon, unique to humans, is called joint intentionality.

Great ape communities do not operate this way. On the one hand, it is true that they are aware of the intentions of others. For example, they may anticipate the choices a competitor will make in a foraging strategy, so that they can appropriately adjust their actions. However, even though they have some understanding of the mindset of others, they seem unable to picture the other as a collaborative partner—only as a competitor. There are various ways in which investigators have determined this, but the following experiment illustrates the difference. If an investigator points to a bit of food, the ape will visually follow the pointing action and go to the food. However, if there are two buckets—one that has food and one that does not—when the person points to the one that has food, an ape cannot grasp that the intent of the pointing action is to help her know which bucket has the food. Apes do not understand that others may have a desire to help. This is night-and-day different from the predisposition of humans. Children as young as twelve months can successfully play a "let's find the hidden toy" game; they are able to understand the intent of an adult pointing to one of two objects, either one of which could conceivably be concealing a hidden toy.[8]

Since all four of the great ape species—orangutans, gorillas, chimpanzees, and bonobos—lack the ability to understand a cooperative mindset, this implies that the lack of this skill is the primitive mindset. What is new in the lineage then is our joint intentionality, and the last common ancestor (LCA) would not have had the ability to picture a helping intent in the mind of others. Tomasello summarizes it this way:

> Overall, as paradoxical as it may sound, our best guess is that LCA individuals had rich social lives with long-lasting relationships, but—as compared with humans—their sociality was still somewhat individualistic. When hunting, they could not put their heads together with others to form the shared goal of working together, and they had no tendency to share resources fairly among all relevant parties. Chimpanzees and bonobos,

7. Tomasello, *Natural History of Human Morality*, 40.

8. Behne et al., "Twelve Month Olds' Comprehension."

and so the LCA, are and were very social, but only in a kind of instrumental way.[9]

GENETIC EVIDENCE FOR COMMON ANCESTRY OF HUMANS AND GREAT APES

One of the chief themes of this book is the question of how the human capacity to be so remarkably skilled at and uniquely attuned for functioning as cooperating units came about. The answer is grounded in the notion that the genes within our lineage have been gradually changing over the several million years since our species shared a common ancestor with chimpanzees. Is this presumption of common descent true, however? How can we know that our gene-set and that of the chimpanzee are each derived from the same set of genes found in a single ancestral species from whom we are both descended? A whole book could be written about this and indeed one especially powerful book has been written by New Zealand geneticist and Christian leader Graeme Finlay.[10] For an intermediate-depth analysis see the article, "Testing Common Ancestry: It's All About the Mutations," by Stephen Shaffner on the BioLogos website.[11] Here I will summarize just one component of the evidence, but this one demonstrates especially poignantly how we know that the two gene pools are derived from one—that of our common ancestor (the LCA) so long ago. Although the focus here is on genes, there is a vast array of thousands of hominin[12] fossils—many dated quite precisely—which also demonstrate that creation of our species has occurred through a long lineage of gradual change in skeletal features.[13] The skeleton, though, is a product of genetic instructions contained in DNA, and it is at the DNA level where the evidence is direct.

You came into being, of course, when the genetic information from your father fused into a single package with that from your mother. Because of this, if you analyze *your* genetic information (your DNA) half of it would match your mother's, and the other half would be a match to that of your father. From each parent, you carry 3.1 billion bits of code (or bases of

9. Tomasello, *Becoming Human,* 14.

10. Finlay, *Human Evolution.*

11. https://biologos.org/articles/testing-common-ancestry-its-all-about-the-mutations .

12. Recall that the term "hominin" is the word used to refer to all species in our lineage that resemble humans more than they resemble apes.

13. Tattersall, *Masters of the Planet.*

DNA)—6.2 billion altogether—and they are almost a perfect match-up of one parent or the other, as I will explain.

Notice I said *almost* perfect. The match-up is not quite exact. On average, of the 6.2 billion units of code that are used to build your body, we would expect that about seventy-seven of those 6.2 billion bits would have changed from that of your parents. To *make* the sperm and egg that give rise to you, the DNA of your parents was copied. Although the copying mechanism is almost perfect, there are still an average of seventy-seven errors made in each generation. The errors are called mutations and the rate at which they occur has been studied extensively. The average rate is remarkably similar over time.[14]

What this means is that if we go back two generations, you would have about 154 differences unique to you, not present in the DNA of any of your four grandparents. And there would be 231 differences, not found in the DNA of any of your eight great-grandparents.

Now that is not going back very far—just three generations—but already there are 231 changes in the code (the bases of DNA) that are now unique to you and not present in any of your great-grandparents. If we go back ten generations (that would be about the year 1750), there would be about 1,024 different people from whom you are genealogically descended. Of your 6,200,000,000 DNA bases (units of code), almost all of it would have been passed down unchanged from those ancestors. But not quite all, of course. Approximately 770 bases would have changed, so that leaves about 6,199,999,230 (99.99998 percent) unchanged.

Much work has been done to check this mutation rate over time and there is very strong evidence that shows the average remains quite constant, varying not more than twofold.

If, indeed, humankind arose through descent from an ancestor that we share with chimpanzees, there would have been mutations occurring all the way back through time. Copy errors occur, we know how often, and the hypothesis we are testing predicts that rate should hold true all the way back to the common ancestor we share with chimpanzees. So how many changes would we predict would have taken place if we really did each share the same set of great, great, great, great . . . grandparents so long ago?

First, we need to estimate the number of generations that have existed in the time since the hypothetical shared great, great, great, great . . . grandparents would have lived. Looking at the fossil record, as we go back deeper into time, we find fossils with bodies looking less and less like ours and

14. The rate of mutations is greater in older fathers, but the *average* is consistently about seventy-seven mutations per generation. See Kong et al., "Rate of *De Novo* Mutations."

more and more like the hypothetical ancestors of the lineage we supposedly share with chimpanzees. In examining this change in the fossils, paleoanthropologists put the date of our last common ancestor with great apes as recently as five million years; others place it further back, a few as far back as ten million years. For the sake of illustration, let's say five million years and let's estimate a generation time of twenty years. That is about 250,000 generations. Rather than examining all 6.2 billion bases (coding units) in the genome, we will analyze a representative sample. In one study the sequence of 1,944,102 units of code was compared in humans and chimpanzees.[15] The researchers found that there were 24,165 differences between the two species. This percentage (1.24 percent) is like what we find for single base differences between chimpanzees and humans in the 6.2 billion bases in the whole genome. So, it is fair to say that this is likely a representative sample.

So, what is the number of changes that would be *predicted* if there were seventy-seven changes per generation over 250,000 generations? The answer is 13,181. That is assuming five million years since the last common ancestor. If the last common ancestor lived ten million years ago the number of generations would be 500,000 and the number of expected differences would be 26,362. Those are the predictions if we go back through each of our lineages (chimpanzees and humans) over those two specific time spans. These numbers compare very favorably to the 24,165 observed differences, a number which is within the expected range. The point, however, is not to get a precise time of the common ancestor; after all, there are approximations in the calculation. (Twenty years per generation is one approximation.) The bottom line is that the number of coding differences between chimpanzees and humans is extremely close to what one would expect given that we know how many mutations are introduced each generation by direct measurement in today's families. The passage of between 250,000 to 500,000 generations provides the precise amount of time needed to account for the genetic difference between the chimpanzee and humans.

If this were the only consideration, it would come close to being a mathematical proof of common descent. However, there is another even more powerful consideration.

As we have been discussing, DNA exists as a sort of code consisting—for the human body—of 6.2 billion coding units arranged in a highly specific sequence. The units of the code are simple—just four different molecular components: Adenine (A), Guanine (G), Thymine (T) and Cytosine

15. See Jobling et al., *Human Evolutionary Genetics*, 54; Ebersberger et al., "Genome-Wide Comparisons," 1490; Nachman and Crowell, "Estimate of Mutation Rate per Nucleotide," 297; this calculation includes an estimate of the mean effective population size (10,000).

(C). Following the meeting of sperm and egg that gave rise to each of us, the code is transcribed and translated over the following nine months as our being is transformed from a single microscopic cell (the fertilized egg) into a fully formed unique human body. To demonstrate my next point, consider the following stretch of DNA code.

AAGTAATTCAGTCAATTAA**C**GGGTAACACTTTA

As we proceed, we are going to focus our attention on the C that is next to a G, the one underlined and in bold font. Notice there are four other C's in this sequence, but the one that is next to a G is special in the way I am about to describe. We know from many experiments that when a C is to the immediate left of a G, it changes (or mutates) at a much higher frequency than when it is next to a T, A, or another C. Indeed, by analyzing the DNA coding sequence of a child relative to her parents, such C's mutate at a frequency that is eighteen times higher than when a C is next to any of the other three bases. We know the reason for this.[16] For our purposes, however, it is sufficient to note that this huge spike in mutation rate at spots with this characteristic is caused by an inefficiency in the cell's repair process at such sites on the DNA.[17] The copying errors cannot be repaired efficiently at such locations. The result in study after study is that when a C has a G on its right-hand side (as we have drawn it here) the probability of that C mutating in a heritable fashion is much, much higher.

With that as background we can make a prediction. Since C's will mutate eighteen times faster if they have a G on their right-hand side than if any of the other three bases are in that position, it implies that humans and chimpanzees would not only differ by a certain prescribed amount as we showed above, but in cases where a C was next to a G they would differ *eighteen times more* frequently.

So that is the prediction, but does it hold? Investigators analyzed ~2,000,000 units of code and compared them unit by unit in both chimpanzees and humans. What they found was that when a C was next to an A, T, or another C there was an average of 376 differences. If, on the other hand, the C had been next to a G there was a huge spike—6,770 cases where chimpanzees and humans differed. And 376 vs. 6770 is an 18.0-fold increase in

16. Holliday and Grigg, "DNA Methylation and Mutation," 61; also, see Jobling et al., *Human Evolutionary Genetics*, 55–57.

17. When DNA is copied in the production of sperm and eggs, the cells have a way of detecting and repairing most copying errors. Typically, as mentioned above, cells leave an average of seventy-seven changes unrepaired. A disproportionate number of these unrepaired changes are in spots where a C is next to a G.

the amount of difference between chimpanzees and humans where a C had resided next to a G. This, of course, is the expected eighteen-fold difference. If chimpanzees and humans had a common ancestor and if mutations had been occurring at a rate and pattern that we can measure through lab experiments today, then it accounts in a very precise way for the coding differences between humans and chimpanzees.[18]

These two sets of observations, together with the gradual change in the fossil record over time using dating techniques that have been verified repeatedly, provide strong evidence that God has created humankind through a gradual process of descent from an ancestor that we share with chimpanzees. Again, however, I emphasize that for a remarkably consistent book-length demonstration of other aspects of the genetic data, see Graeme Finlay's book.

As mentioned above, the gradual change in hominin fossils through time is a direct consequence of the changes in the DNA. Genes are housed in DNA, so DNA changes correspond to gene changes, which in turn correlates with skeletal changes (along with other changes) that, through time, come to take on characteristics that increasingly resemble features of our own skeletons. From walking on four legs to two, from using forelimbs for swinging through trees to using them for the finely tuned manipulation of objects, from small brain cavities to large ones, and from the faces of an ape to the facial features of a human being, all these changes show up in the fossil record, all are a product of gene changes, and all are primarily restricted to the continent of Africa.

THE HUMAN LINEAGE

It is one thing to identify transitional features in the fossil record, but it is another to show that their existence was timed in a manner consistent with a progressive sequence from ape to human. Nowhere is the progressive timing of the transitions better documented than in the Great Rift Valley of northeastern Africa where Ethiopia, Kenya, and Tanzania exist today. This valley has come into existence through the geological splitting action caused by two continental plates that have been sliding apart from one another for millions of years. Still today, the resulting valley is subject to flooding just as it has been throughout this span—animals still get stuck in the muddy

18. The scope of this book limits this discussion. One other compelling piece of genetic evidence is outlined in three short videos on my YouTube channel, "Coming to Peace with Science." To access, you can Google: YouTube, Falk, Chromosome 2 or go directly to: https://youtu.be/40QbU4O8--E, https://youtu.be/GdLGCrbQJQU, and https://youtu.be/HAvqDl3yLDs.

flood sediment just as they have for millions of years. After their bodies decay within that mud, the resulting skeletal remains—if undisturbed—are cemented in place. As the sediment continues to harden over future millennia, the animals buried there become tomorrow's fossils. It is possible to determine the age of "yesterday's" fossils due to another unique geological feature of the region. The geological instability caused by those continental plates sliding apart from one another has resulted in frequent volcanic eruptions over those several million years. Whenever a major eruption occurred, a layer of ash was laid down over the entire landscape. Hence, like a multi-decked sandwich, each layer of sediment lays on top of one layer of ash and under a different (younger) one. The fossils themselves do not have minerals suitable for determining age, but the ash layers do. So, by dating the ash layers above and below a hominin fossil, we can get a fairly accurate estimate of the fossil's age.

The oldest mostly complete hominin fossil in this region is *Ardipithicus ramidus*, nicknamed Ardi. It is about 4.4 million years old. The story of finding Ardi begins in November 1994, when a single hand bone was found peeking out of some sedimentary rock. Careful excavation of the site resulted in the recovery of 45 percent of the skeletal remains from one individual, a female, who, based on her skeletal features, would have been bipedal—able to walk on two legs much like us. Her features imply a lifestyle adapted to both swinging through the trees and walking on the ground.[19] Like today's great apes, she had a splayed big toe sticking out sideways that would have served well in tree climbing. And her hands and arms would have been well suited for life in the trees as well. No stone tools have ever been found at sites of this age (or even up to one million years later) so we can presume that the techniques for making stone tools had not yet been invented.[20]

Although *Ardipithicus* is the most complete early hominin fossil found to date, there are several other specimens, less complete and not discussed here, that are a million or so years older. What is especially important to note is that the first sign of hominin fossils dates to about five to six million years ago. This is only the last 0.1 percent of earth's history. No hominin fossils have ever been found in rocks older than this even though that period represents 99.9 percent of earth's history. Furthermore, despite geological surveys the world over, no hominin fossil—other than those representing our own genus,[21] *Homo*—has ever been found on any other continent than

19. Lovejoy et al., "Combining prehension and propulsion," 72.

20. Harmand et al., "3.3-million-year-old stone tools," 310.

21. "Genus" is a reference to a term used in classifying organisms. We are members of the genus *Homo* and the species *sapiens*.

Africa. Thus, paleontologists conclude that our lineage arose at a very specific time in earth's history and on a single continent—Africa.

Although the fossil record is sparse prior to 3.8 million years ago, the situation changes dramatically after that. In studies covering the next two million years, thousands of fossils from hundreds of individuals representing almost a dozen species have been found. The genus *Australopithicus* predominates throughout this period and *Australopithicus afarensis (A. afarensis)*, one of five species, is the most thoroughly represented and studied. Investigators have found several almost complete specimens of *A. afarensis* (the well-known Lucy is one) spanning the interval of 3.8 to 2.9 million years ago (hereafter MYA), after which the species is never found again. The shinbone of *A. afarensis* was structured such that it would have attached to the ankle in a manner that resembled how ours is attached rather than the way it is joined in the chimpanzee. Moreover, the structure of the ankle is humanlike. On the other hand, the face was apelike, with a flattened nose and strongly protruding lower jaw. The brain was about one third the size of ours. The shoulder blade was not humanlike; rather, it resembled that of a gorilla. In short, *A. afarensis* had a body that still would have been well adapted for an existence in trees as well as the ground (just like the somewhat different body characteristics of *Ardipithicus* a million years earlier). Interestingly, the hyoid bone, a part of the voice box, was structured in a manner that is gorilla-like and this suggests Lucy and her kin had apelike vocal abilities.[22]

Although the anatomical features of *Ardipithicus* strongly suggest that walking on two feet arose early in hominin history—at least 4.4 MYA—the most poignant demonstration of bipedality comes from a set of 3.6-million-year-old footprints of two individuals extending over about eighty feet in what would have been wet volcanic ash. The big toe was almost parallel to the other toes; it was not splayed outward as with Ardi. Detailed analysis of the gait of the pair indicates that they walked in a manner almost indistinguishable from us. The footprints likely belonged to Lucy's species, especially given that fossils of her species (*A. afarensis)* were found nearby in the same layer of volcanic ash.

Beginning about 2.5 MYA, representatives of another genus closely related to *Australopithecus* made its appearance. Individuals of this genus, *Paranthropis* (represented by three species), had larger, tougher teeth, a bigger area on the skull for attachment of huge chewing muscles, and a large face. Since the *Paranthropis* species have several new anatomical features

22. Tattersall, *Strange Case*. loc. 2911.

not found in our lineage, most investigators believe they are a side branch, not on the direct lineage to *Homo*. By 1.2 MYA, all were extinct.

Our genus, *Homo,* enters the fossil record with a whimper, not a bang. All we have so far from 2.8 MYA to 1.9 MYA are two independent finds from Ethiopia—portions of the *Homo* jaw—one from a 2.8-million-year-old site[23] and the other dated at 2.3 MYA.[24] The whimper changes dramatically at 1.9 MYA, when *Homo* begins to predominate the hominin fossil record. As *Homo* emerges, *Australopithicus* fades from the fossil scene (the most recent find is likely at least 1.75 million years old). *Paranthropis,* as mentioned above, persists as a side branch up to 1.2 MYA.

Our genus, *Homo,* is generally defined by a larger brain, a narrower, less stocky body shape, shorter arms, and longer legs. It is not currently possible to define a specific trajectory to the various *Homo* species, in part because it seems likely that the genus originated well before the fossils became abundant. If further fossil finds confirm and extend the finding of a *Homo* jaw at 2.8 MYA, then that would mean we currently have a 900,000-year period where *Homo* existed but almost no fossils have been found.

As *Homo* becomes well established in the fossil record beginning at 1.9 MYA, two species, *Homo habilis* and *Homo rudolfensis,* stand out as being the most primitive (that is, *Australopithicus*-like). The somewhat more "derived" (less primitive) *Homo erectus*[25] appears at other locations at just about the same time. It seems likely then that there were as many as three lineages of *Homo* present in Africa during this period, along with some species of *Australopithicus* and *Paranthropis* that had not yet gone extinct.

Almost simultaneous with the appearance of *Homo erectus* in Africa's fossil beds, the same species (or a closely related variety of it) emerges onto the fossil scene in Asia. Since no hominin fossils older than this have ever been found in Asia or Europe, it seems likely that some members of the *Homo erectus* species left Africa about this time and journeyed northward. The earliest finds—five skulls in what is now the Republic of Georgia—date back to at least 1.85 million years. However, there is a 1.6-million-year-old *Homo erectus* site near Beijing in China and another one younger than that in Indonesia. Indeed, *H. erectus* (or closely related varieties) persisted in parts of Asia until at least a few hundred thousand years ago and probably even more recently than that. So, the geographic distribution of *Homo*

23. https://www.sciencemag.org/news/2015/03/fossil-pushes-back-human-origins-400000-years.

24. Kimbel et al., "Systematic assessment of a maxilla," 235.

25. There is some disagreement among paleoanthropologists about the nomenclature of the various *Homo* specimens within this timeframe. This need not concern us here.

erectus—from various locations in Africa to many parts of Asia—was widespread. Throughout the first million years of the existence of this species, the brain size was gradually increasing. Most of the earliest recovered skulls indicate a brain size only slightly larger than that of apes. However, over time there was a gradual increase in brain volume until it eventually reached a volume close to that of our own.

In 2013, the largest and most complete set of hominin fossils ever found were discovered in two extremely difficult-to-reach caves in South Africa.[26] The fossilized skeletons represented a new species, *Homo naledi.* On the one hand, this species looked comparatively primitive. It had a very small brain and it had an archaic-looking pelvis, shoulder, and rib cage that bore considerable resemblance to *Australopithicus.* Interestingly, though, the specimens were dated to about 250,000 years ago, more than 1.5 million years after *Australopithicus* had likely gone extinct. Just as puzzling, though, is that its hand, wrist, foot, and ankle were all quite like our own and because of that it has been placed in our genus, not *Australopithicus.* We will have more to say about this fascinating species a bit later.

INCREASING BRAIN SIZE

Except for these *Homo naledi* individuals, the most remarkable feature characterizing the changes in our genus through time was the increase in brain size. In the words of Vanderbilt neuroanatomist Suzana Herculao-Houzel:

> One of the hallmarks and most remarkable aspects of human evolution is that the brain of *Homo* species increased very much and extraordinarily rapidly—nearly tripled in size during the last 1.5 million years, really—while the brains of our great ape cousins stagnated at the same size they have to this day. To put this into a wider perspective, it took around 50 million years for primate brains to go from micro lemur sized to gorilla sized (an increase of 29 billion neurons), but only 1.5 million years to add another 57 billion to *Homo* alone—nearly twice more neurons. Compared to that of great apes, our evolutionary history indeed seems extraordinary: an oddity that singles us out.[27]

The following figure illustrates this graphically:

26. See Berger, *Almost Human.*

27. Herculao-Houzel, *Human Advantage,* 187.

FIGURE 2.2

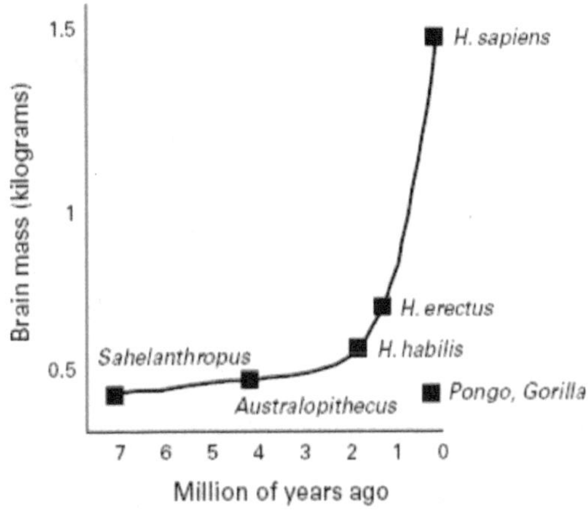

Change in Hominin Brain Size Through Time[28]

So, what specifically happened that enabled this dramatic growth in the number of neurons? Just as gene changes specified the changes to skeletons, it would have been changes in the DNA that instigated the enhanced brain growth. Since this was so important to the origin of our species, I am going to summarize what we currently know about the nature of these genetic changes. Much could be written about this, but we can accomplish our purposes in just a couple of pages without getting particularly technical.

First, we need to say a bit more about DNA. As mentioned above, DNA is a molecule found inside of almost all our cells. It contains a code with the instructions to produce the human body and keep it functioning. The entire code consists of 3.1 billion units present in two copies, one from each parent, making 6.2 billion altogether. DNA is packed into discrete packages called chromosomes. We have twenty-three different chromosomes, each with a different DNA molecule. Each type of chromosome is present in two copies (again, one from each parent), making forty-six altogether. Each gene is a tiny section of a DNA molecule that makes up the chromosome. In general, one gene provides the information on how to build one specific protein molecule (or part thereof). We have about 21,300 genes distributed in the DNA molecules packed into forty-six chromosomes—with two versions of each. Each chromosome (DNA molecule) carries on average about one thousand genes or, in other words, the coding instructions for making about one thousand different protein molecules. Each human being starts

28. Herculao-Houzel, *Human Advantage*, Figure 11.1, used with permission.

life as just one cell—the fertilized egg with its forty-six chromosomes. Soon thereafter, however, the one cell divides in two, then four and so on into adulthood. As the cells duplicate dividing into two cells, the chromosomes within the cells (having been replicated themselves) are also distributed evenly into the two daughter cells.

Let's picture a tiny portion of a typical human chromosome. We will focus on just four of its thousand genes, but they all exist together lined up side-by-side:

Gene 1—Gene 2—Gene 3—Gene 4 etc.——————————————Gene 1000

These genes (nothing more than sections of a DNA molecule) must be duplicated if they are going to be passed into the two daughter cells of a developing embryo.

For Daughter Cell #1:
Gene 1—Gene 2—Gene 3—Gene 4 etc.——————————————Gene 1000

For Daughter Cell #2:
Gene 1—Gene 2—Gene 3—Gene 4 etc.——————————————Gene 1000

Sometimes, when a chromosome is replicated, a mistake is made, in which a sort of "stutter" occurs so that a portion of the DNA molecule is present in a duplicated condition side-by-side, as shown below for our hypothetical Gene 3.

Gene 1—Gene 2—Gene 3—Gene 3^{dup}—Gene 4 etc.————————Gene 1000

In effect, when an error of this sort occurs it simply creates an extra copy of the original gene. As we proceed, we are going to follow how that extra copy of a particular gene became important in increasing the brain size of our lineage relative to that of our great ape cousins. We have arbitrarily called the original unduplicated version Gene 3, although it does have an official name.[29] We are calling the duplicated spare Gene 3^{dup}. It resides alongside the original and is a faithful copy of it, except for being a little shorter, missing the coding information for the tail end of the protein product.

29. The official name of Gene 3 is ARHGAP11a.

In chimpanzees, bonobos, and gorillas, the genes are lined up like this; the same genes are present, but they do not carry the duplication:

Gene 1—Gene 2—Gene 3—Gene 4 etc.————————————Gene 1000

What this means is that at some point in our ancestral lineage, a duplicating error occurred, and ever since, this segment of the chromosome with its two side-by-side copies of Gene 3 has been faithfully passed from parents to offspring so that today every one of us has that extra copy of Gene 3 in this chromosome. We know how long ago the duplication occurred; it came into existence about 5.2 million years ago. Analysis of the DNA of Neanderthals shows that they, like us, but not like the great apes, had the duplicated copy of Gene 3. This is not surprising because they are our cousin species; we share the same lineage, the same ancestors up until about 600,000 years ago, and the duplication event occurred millions of years earlier.

At some point, we do not know exactly when, there was a tiny change in the DNA code of Gene 3^{dup}. Recall that DNA contains four code letters (or bases): A, G, C, and T. A typical gene contains several thousand bases. In the case of Gene 3^{dup}, one of the several hundred C's was changed to a G. That single C-to-G coding change altered the gene's coding instructions and significantly altered the protein product of this gene.[30] We'll designate this altered copy of the duplication, Gene 3^{dup*} and we can diagram this section of the chromosome like this:[31]

Gene 1—Gene 2—Gene 3—Gene 3^{dup*}—Gene 4 etc.———————Gene 1000

All humans have this version; indeed, even our Neanderthal cousins had it. Remarkably, this single change (one base out of the 3.1 billion)—a C to a G in this one gene—significantly impacts the number of nerve cells

30. The technical details of what has happened at the molecular level are beyond the scope of this book. For those readers who have a biology background, however, here is a bit more detail. The single base change in Gene 3^{dup} alters how the precursor RNA of the gene is spliced as it matures into mRNA. The change in splicing removes fifty-five bases that would otherwise have remained in the mRNA. Because of the new boundaries of the splice, there is a shift in the reading frame of the mRNA, resulting in a much different sequence of amino acids at the tail end of the protein product. Those extra amino acids give the protein a new set of properties.

31. The official name for Gene 3^{dup*} is ARHGAP11B. The key technical reference for this story is Florio et al., "Single Splice Site Mutation." For a more general summary, this video is great, albeit still a bit technical: https://carta.anthropogeny.org/mediaplayer/play/309473/301958.

(neurons) in the brain. Although for ethical reasons this cannot be demonstrated directly through experiments with humans, it *can* be shown through insertion of the human Gene 3^{dup*} into the embryos of mice. With that extra gene, mouse embryonic brain stem cells proliferate more frequently and produce more neurons. Furthermore, the brains of these genetically altered mice frequently fold in a manner that resembles the highly folded nature (i.e., the gyrification) of the human brain. Changing the gene back to its original state demonstrated that the single coding change of a C to a G was responsible for the higher proliferation. If this corrected Gene 3^{dup} is placed into the mouse embryo instead of the mutated Gene 3^{dup*}, it has no effect on mouse brain development. Thus, the increase in brain stem cell proliferation only occurs if that one code letter is a G. If it is a C there is no enhancement of the mouse's developing brain. To test these results further, the investigators repeated the experiments using ferret embryos, and then repeated it again more expansively on the embryos of marmoset monkeys; the results were the same.

The biological details of how the protein product of Gene 3^{dup*} causes the brain cells to proliferate more often is fascinating, albeit largely beyond the scope of our story here. Suffice it to say that the protein specified by this new, human-specific gene results in an increased energy supply for the embryonic brain stem cells, and this boost powers an increased level of cell division.[32]

It is not possible to pinpoint when this mutation occurred other than that it was after the duplication event 5.2 million years ago, and prior to when our lineage separated from that of Neanderthals about 600,000 years ago. So even though we cannot pinpoint a date for the mutation, given the dramatic effect that this gene has on the number of neurons, this mutation was likely one of the key genetic changes responsible for the brain expansion in our lineage. According to the fossil record, this began in earnest about two million years ago. There were undoubtedly many other genetic changes

32. Here is a little more detail for readers with a biology background. This single base change—the C to a G—results in a protein with a new amino acid sequence at one end. As a result of that change, the protein is targeted to enter the mitochondria of the brain stem cells. In the mitochondria, this protein interacts in a highly specific fashion with a protein that controls the amount of calcium entering the mitochondria. The change in calcium concentration results in a metabolic change which increases the activity of the Krebs cycle. And as you will remember from biology 101, a more active Krebs cycle results in the production of more ATP, which in turn provides the energy to drive an increased amount of cell division. The story of how we know all of this is one of the most beautiful cell biology studies I have ever seen. (See Namba et al., "Human-specific ARHGAP11B Acts in Mitochondria," 867.)

affecting the brain (and we will examine one below), but our knowledge about this one example gives us a sense of how genetic change can facilitate an increasingly sophisticated brain.

Having a more sophisticated, larger brain did not come without a cost to the lifestyle of our ancestors. Herculano-Houzel summarizes the "cost" this way:

> Our single largest advantage over other animals is most easily ascribed to the sheer number of neurons available in the cerebral cortex, and in the prefrontal cortex in particular, to process information in complex, flexible ways that can predict future outcomes and act as required, particularly in a way that intelligently maximizes future possibilities. On the other hand, the human brain does seem special in the amount of energy that it requires. Even though it represents only around 2 percent of body mass, the human brain alone costs some 500 kilocalories per day to operate, a disproportional 25 percent of the daily energy required for the entire human body to work. In comparison, the brains of other vertebrate species cost at most 10 percent of the daily energy budget of the body . . . The source of all this energy is usually glucose from the blood . . . This explains why blood flow to the brain is tightly regulated: to keep the human brain running on blood glucose requires a remarkably constant average of 750 milliliters of blood flowing through the brain per minute, and even a 1 percent drop in flow may cause fainting and loss of consciousness.[33]

Humans have about eighty-six billion neurons and, based on the size of fossilized skulls, it is likely that our lineage has had close to that many for the past 700,000 years or more. That is a lot of neurons and investigators estimate that for a typical primate diet, it would take nine to ten hours of foraging and feeding per day to fuel the high demands of a brain of our size. This might not sound like too much, but great apes start to lose weight whenever food availability requires them to forage more than 7.5 hours per day. So, if our ancestors needed to average nine to ten hours each day they almost certainly would not have thrived and the species likely would have died off. Smaller brains would have won out. Since that is not what happened, how did our ancestors make it? Heculano-Houzel suggests that there would have been four ways that our ancestors could have thrived despite the higher energy costs of a large brain. The first would have been to reduce the size of the body, but the fossil record makes it clear that is not what

33. Herculano-Houzel, *Human Advantage*, 143.

happened. The second would have been to decrease the energy costs of the brain—but ounce for ounce our brains do not use less energy compared to other mammals. The third way of thriving despite the taxing energy demands of a large brain was to spend more time foraging for food—but as indicated above, they were already at their maximum. All of this leaves only one possibility: our ancestors must have been able to increase the amount of usable energy acquired per day.[34] So how was this accomplished?

Interestingly, the brains of our ancestors began to expand at a time which roughly parallels the appearance of stone tools in the archaeological record. Such tools would have aided in hunting game and processing meat. Manufacturing stone tools is a feature unique, of course, to our lineage. True, some animals can use tools—for example, chimpanzees use rocks to crack nuts open and some crow species can even make use of sticks in quite fancy ways to access food. No animal, however, has the foresight to formulate a plan of the sort that would have been required to shape a hand ax out of a stone. Making this tool would have required foresight and reason at a level never before attained in the animal world. As genetic changes enabled the brain to become larger, our ancestors would have needed to procure more energy-rich nutrients without increasing their workload. So, they depended on the extra brainpower associated with increased brain size to accomplish what had previously been impossible. In short, hominin brains were getting larger because there was an advantage to being genetically equipped with more neurons. Such individuals would have been able to more effectively procure the nutrition needed to meet their own increased energy demands as well as the needs of their children's rapidly expanding brains. Thus, individuals with larger brains would leave more offspring, setting up a cascade effect: Larger brains—>more skill in nutrient acquisition—>more offspring surviving to reproductive age.

THE SOCIAL BRAIN

But the larger brains did not just make our ancestors smarter, they also increased their proficiency in working with each other. By "putting their heads together" an even larger mind emerged—"the social brain." Our ancestors did not make their stone tools in a social vacuum; they taught each other this sophisticated skill. And they didn't go off on a hunt by themselves; they planned together and developed cooperative strategies. And they didn't just

34. Herculano-Houzel, *Human Advantage*, 188.

care for their children by themselves; they did so communally. The secret to the success of our lineage was that their increasingly sophisticated brains enabled them to interact with each other in more productive ways—in a manner more attuned to the minds of others. In short, they were becoming communities of socially interacting individuals, and all those neurons not only more efficiently equipped their minds for greater intellectual feats, but they also equipped them for increasingly subtle and unusually efficient means of social interactions, which take brain power too.

Undoubtedly, investigators tell us, the use of fire begins to fit into the story about this time. The formation and long-term use of fire pits leaves a telltale signature that can readily be detected in archaeological field studies. Then, as the size of the brain really started to take off about 1.5 million years ago, fire use became increasingly common. As Richard Wrangham wonderfully documents in his book *Catching Fire: How Cooking Made Us Human,* the significance of fire cannot be overstated. Cooking food, instead of eating it raw, made calories accessible in a manner that changed everything. Cooking uses heat to denature proteins, break carbohydrate chains, and otherwise modify foodstuffs so that the clumps are smaller, softer, and more chewable. In this way, nutrients become increasingly accessible to enzymatic digestion in the mouth, the stomach, and the intestines. Through cooking, close to 100 percent of calories in meat are made accessible to the body. In contrast, as Herculano-Houzel explains, "the same foods may yield as little as 33 percent of the energy in their chemical bonds when eaten raw because these harder foods are swallowed while still in pieces and thus are broken down and digested only partially."[35]

Figure 2.3 summarizes the corresponding increases in brain volume relative to what we know about hominin activities. The initial stage corresponds to the origin of tool manufacture; bipedality freed up the hands for the dexterity needed for making tools to assist with hunting and gathering. During that stage, there was modest brain growth. Then with the advent of fire and its concomitant increase in caloric efficiency, there was a dramatic increase in brain size in hominin history.

35. Herculano-Houzel, *Human Advantage*, 194.

FIGURE 2.3

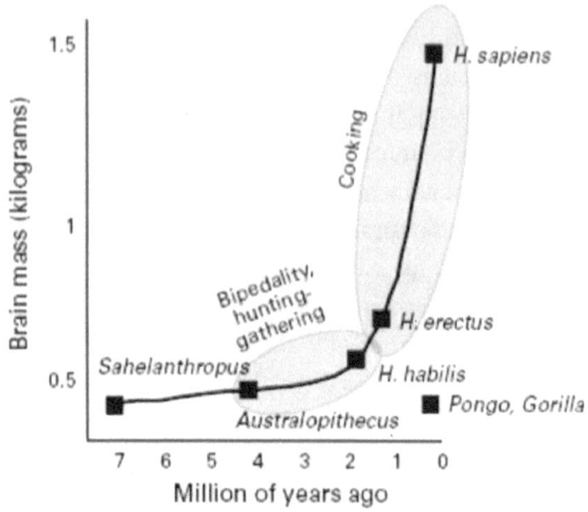

Brain Size and Cultural Activity[36]

Besides the impact of fire on making more calories available to fuel all parts of the body at lower cost, there is a byproduct to cooking over a fire that is likely almost as important as the release of nutrients. As we all know from our own camping experiences, fire fosters community. It nurtures a spirit of working and living together as a unit—not as solitary individuals—but as bonded persons who love and care for one another. So just as fire provided the fuel for the expanding brains of our ancestors, it also fueled the development of a social mindset. This, in turn, would have been facilitated by more neurons as we continued to become increasingly attuned to finer and finer social cues.

As the minds of our ancestors became more aware of all that surrounded them—both in their natural environment and in their community—they may have started down a path of increasing awareness of the reality of the nonmaterial world as well. I will have much more to say about this as we proceed, but for now, suffice it to say that as evolving hominins were becoming more sensitive to others in their communities (who they can see and whose influences they can assess directly) they might also have developed a sense of an external reality. As Christians who live into and experience the love of God today, let's think about how the God we experience would have become engaged with this growing perception of an external reality in those

36. From Herculano-Houzel, *Human Advantage*, Figure 11.2, used with permission.

early hominin communities. Of course, all that exists was brought into existence by God—God is never outside the process. Might we wonder though, if there would have been a new dimension to God's activity once members of these communities began to consciously reach out for the divine? If so, might we expect that some communities would be more attuned to listening to the "voice" of God's goodness and God's love than others? Certainly, there are community differences like that today; might it not have been true in those early days as well? If so, any communities attuned to God's presence, however "childlike,"[37] would likely have functioned better. Love, after all, makes a difference. Such communities may have been more effective, for example, at carrying out joint tasks than those communities which paid no heed to this "call" for goodness. If so, could a positive response to this divine-based urging have been a sort of selective force, enabling responsive populations to thrive and unresponsive ones to diminish? Hold that thought; we will have more to say about it as we proceed.

INCREASING CONNECTIVITY

The increase in brain size was taking place, of course, through the genetic changes that facilitate an increase in neuron number during the growth and maturation of the brain in fetal development and on into childhood. As a way of illustrating how the brains of our ancestral species became larger, thereby making this possible, we discussed one genetic change that likely was an important key to increasing neuron number and brain size. There is more to the story, though. An increase in neuron number is not the only significant change. The number of connections between those neurons increased as well. How did this increase in connectivity come about?

As mentioned above, humans have about eighty-six billion neurons. Each of those neurons communicate with countless others through junctions called synapses. In essence, the neurons "talk" to each other through the chemical and electrical signals they send through connections to certain neighboring neurons—those equipped to "hear" what they are saying. Again, we are staying away from technicalities, but it is possible to discuss the basics without getting buried in detail. The increased level of communication between neurons in our lineage was facilitated, not surprisingly, by gene changes. We are going to look at just one of these—a particular change that took place 2.4 million years ago, a long time before our species, *Homo sapiens*, came into existence. Just like the change in Gene 3 that we

37. I use the term "childlike" here in the same sense that I think Jesus did—an uncomplicated faith accompanied by an unadulterated desire for goodness.

discussed above, this change also involved the duplication of a preexisting gene. We will call this one Gene 4, to avoid the technical distraction of its formal name.[38] Gene 4 is an ancient gene; it is found in all mammals. It provides instructions that play a key role in establishing those connections between neurons. I have already mentioned that neurons use chemical signals to "talk" to each other. Here is a little analogy to give you a sense of how this works. We can think of neurons as having little mouths[39] through which they speak to other neurons. From those little "mouths" they release chemicals[40] that serve as words in communicating with neighboring neurons. The neurons to which they speak have tiny receivers—little "ears"[41]—receptors through which they "hear" the chemical signals. Gene 4 has a highly specific role in establishing connectivity between neurons. It contains the plans for making a protein that regulates the formation of these little "ears," hence, it ensures the efficiency of crosstalk as parts of the brain communicate with each other. More efficient connectivity between neurons leads to increased cognition, which reaches its zenith in the hominin lineage.

All tested mammals have Gene 4. About 2.4 million years ago, in our lineage a portion of Gene 4 was duplicated, and the extra truncated copy was inserted at a different chromosomal position. In what follows, I will refer to this extra, partially duplicated copy as Gene 4^{dup*}.[42] Remember what genes do—they provide the instructions on how to make a protein. So, because of the duplication, there are now two versions of the protein product of Gene 4 produced in developing fetal and childhood brains. Since Gene 4^{dup*} is a duplicate of only part of the original, it provides the plans for making only part of the protein product. Interestingly, the two versions of this protein (we will call them Protein 4 and Protein 4*) stick to each other. Since they adhere, the neurons now contain a Protein 4-Protein 4* complex (termed a dimer). Unlike the situation when only Protein 4 is present, under these conditions the 4-4* dimers delay the production of neuronal projections that contain the little "ears." For reasons not yet understood, the delay in producing these projections from the neuron results in those same projections protruding further and being more numerous.[43] Thus, because of the

38. The official name for Gene 4 Is SRGAP2.

39. Known as synaptic vesicles.

40. Known as neurotransmitters.

41. Known as receptor molecules.

42. There were actually three duplications that occurred at separate times; however, this is the only one relevant to the current discussion. Its official name is SRGAP2c. See Dennis et al., "Evolution of Human-specific Neural SRGAP2 Genes," 912.

43. Charrier et al., "Inhibition of SRGAP2 Function," 923. For a nice general summary of this see: https://carta.anthropogeny.org/mediaplayer/play/309475/30196.

slowdown in the formation of connections, ultimately there are significantly more of them. With that, we reach the bottom line—the partial duplication of Gene 4 slowed down the maturation of brain connections, but in the end the connectivity was more efficient. What this slowdown meant is that compared to great apes, brain maturation was delayed into childhood and the experiences of childhood could influence brain-wiring. This is different than great apes, where more of the wiring of the brain is established in the womb, and less is tuned by social experiences. Furthermore, the existence of Protein 4* resulted in greater connectivity such that neuronal interactions in the brain became more effective. The significance of this change cannot be overstated. In humans, childhood extends much longer compared to great apes. Indeed, the brain of a child is still growing substantially at birth and this is one reason why the skull of babies has not yet closed over at this time—there needs to be room for brain expansion.

In our discussion of Gene 3, we noted the experiments in which the brains of mice developed a larger number of neurons and frequently a humanlike folding in their brains when provided with the human version of that gene's partial duplication. Similar mouse experiments have been done with Gene 4. Remarkably, the formation of the dimer between the mouse's Protein 4 and the human Protein 4* (produced because of the presence of human Gene 4^{dup*} in one of its chromosomes) increases the quality of connectivity in the mouse brain. If mice have the Gene 4^{dup*} from humans, not only is there a delay in the formation of the projections on which the "ears" are situated, ultimately there are more of them. But the most amazing thing of all is that mice that carry this extra gene are cognitively more adept at carrying out a particular task compared to mice that only carry their own version of the gene. Having both gene products present not only mimics this aspect of the connectivity between neurons, but it also improves cognition.[44] This implies then that within our own lineage 2.4 million years ago, this duplication was another key genetic change that allowed individuals to thrive. Certainly, there were other genetic changes, likely many, which also influenced this process—but studies with Gene 4 have provided insight into the sort of thing to expect as investigations continue.

SOME THEOLOGICAL REFLECTIONS

Scientific investigation is enabling us to get a little peek into the way in which God has brought our species into being. It did not happen in an instant. There are several reasons why this instantaneous view has long been

44. Schmidt et al., "Human-specific Modifier."

the view of most Christians. By far the most important is that a literalistic reading of Scripture certainly implies that. There are many fine books that lay out the case for alternative ways of reading and understanding the text in a manner that is more consistent with the intent of biblical authors and the way in which God uses Scripture to speak into human cultures across many millennia and countless different cultures.[45] The Bible, as God's word, is not a scientific textbook. In a manner that remarkably transcends time and culture, the Bible speaks to us about the love of God that reaches out to humankind and tells us how—through the presence of God's Spirit—we can be empowered to live into that love. The Bible is a call; it is a call to be holy even as God is holy. It is the story of the process by which God's love became known in a highly specific way to humankind. That process, just like creation itself, was gradual, extending through many centuries.

In theory, at least, one could propose that Christ should have come into the world much sooner so that more of those earlier civilizations might have been exposed to the love of God exemplified through the coming of Jesus. Instead, God chose to work through a specific group of people as they bumbled their way through the vicissitudes of life in the ancient Near East. It would be inappropriate to question God's wisdom—especially now, millennia later—when every Sunday all around the globe billions of people study the stories of that tiny band of people who lived so long ago. Those stories have become our stories; they speak into our lives today. Be it the freeing of the slaves out of Egypt, the wandering in the wilderness, the crossing into the promised land, the ache of David's sin as we read Psalm 51, or the hope in God that enables us to soar on eagle's wings (Isa 40:31)—life is immeasurably enriched by the fact that God chose to work gradually through the nation of Israel rather than instantaneously. Who are we to think that gradual is less efficient for accomplishing God's purposes than any alternatives that we could dream up? God works through process, even if we prefer quick solutions to the challenges we face.

The way in which God speaks to us about the natural world is different than the way God speaks to us of his love and the call for holy living as exemplified so wondrously in the Bible. Perhaps, we think, it would have been more efficient if God would have just spelled out all the details of the natural world in some sort of instruction manual analogous to the Bible. I am frequently asked why, if God really did create through the evolutionary

45. See John Walton's Lost World series, for example: *Lost World of Genesis One; Lost World of Adam and Eve; Lost World of Scripture; Lost World of the Flood.* Other fine books include: J. Richard Middleton's *Liberating Image;* Scot McKnight and Dennis Venema's *Adam and the Genome;* Iain Provan's *Seriously Dangerous Religion;* and N. T. Wright's *Surprised by Scripture.*

process, doesn't the Bible just tell it like it really was. As I have heard biblical scholar John Walton state so often, "The Bible was not written *to* us, it was written *for* us." It was written *to* the communities in much different times. They were asking very different questions than we ask today. Neighboring nations had their creation stories too, and the people of Israel needed one that answered *their* "why" questions about creation, not *our* "how" questions. However, as we picture God bending down over the lifeless body of Adam[46] to breath his life into Adam's being, God is telling us about our purpose too. The Creator of the entire cosmos described so majestically in Genesis 1 bends down from on high and breathes his infinite life into the finite body of Adam. Adam is an archetypal representative of us all, John Walton and others tell us. We are all loved and each of our lives matter.

This does not mean the questions we ask about the natural world today are not legitimate or important. The natural world, after all, is *God's* creation. Just as we learn about God's character by studying God's Word, so we learn more about God's nature by studying God's works. Those I interact with who want to know what is important to me can listen to what I say, but for the whole picture, it is my actions, not just my words, which provide the most complete representation of what kind of person I am. Therefore, the mode by which Jesus died tells us so much about the character of God. But the study of nature is also the study of God's actions, and we learn about those actions through science.

The scientific process is an amazing gift from God because it allows us to study God's action and interaction with the natural world.[47] It is a gift given specifically to our species: *Homo sapiens.* No other beings in all of creation's history have had access to their Creator's actions in this way. We, alone, are the species that can ask questions about God's creation and obtain answers. Those answers are not written in a book handed down through the generations; they come about by using God's gift to us—our minds. The process of discovery itself is a gift. Ask any scientist about the joy of discovery. Designing and carrying out a set of well-formulated experiments to answer a question about some aspect of the natural world is exhilarating. Working through a series of logical deductions about the meaning of a particular band of light being emitted from a distant star or a not-so-distant planet is a thrilling experience indeed. God has enabled us to experience the joy of discovery for ourselves. And God has also enabled us to experience

46. Hebrew scholars emphasize that the word used for Adam at this point in the text is "mankind" and that the narrative wants to point us to the universality of the story that is being told.

47. We will have more to say about why and how we can think of the natural laws as reflecting the activity of God in chapter 6.

the delight of being able to tell others about the discovery process so that we can share in that joy with each other. Just like the beauty we experience as we spend our lives studying the story of God's love and God's call for holy living in the Bible, so also through using our minds applied to the scientific process we can spend our lives—even if we are not scientists—enjoying the beauty of God's creation.

This chapter has had much to say about God's creation of us—but also provides hints about the character of God. Many Christians believe that God exists outside of time, so, just as 2 Peter 3:8 states, a thousand years are like one day to the Lord, likely we can extend that even further: a billion years is like a day to God. However, God also chooses to live *within* time as God chooses to come down to our level and relate to creation as it exists in time. So, one has to wonder, why did God take hundreds of millions of years from the first multicellular life to bring our species into being or, before that, a couple of billion years to bring about the first multicellular life? We know how much we are loved by God, and being somewhat anthropocentric, we cannot help but wonder why God was not in more of a hurry to get to us. Isn't it all about us, after all?

I think the answer is that in fact it is not all about us. I think Scripture makes it clear that God delights in all of creation and that the goodness of the whole dynamic of creation is an expression of God's love. N. T. Wright's Gifford Lecture Series from 2018 is a wonderful example of studies which are beginning to pay more attention to this point.[48] However, for the purposes of this book, I am emphasizing God's presence in those communities on the African savannas, especially as our ancestors were becoming increasingly cognizant that others had a mind just like their own. As we will see, there are several scholars who suggest that this is about the time that artifacts attributed to religious yearnings appear in the archaeological record. God's love for his creatures did not suddenly begin ten thousand years ago with the advent of agriculture and then civilization. That represents just the last 0.1 to 0.2 percent of the time since the origin of the hominin lineage that led to us. Throughout that time, God had our creation in mind, but God had our ancient ancestors in mind too. And it was God's love for them that began to shape the individuals in those communities, slowly changing their genetic makeup, thereby enabling them to become what we have become today.

48. Wright has also written a book based on these lectures (Wright, *History and Eschatology)*, however, I especially recommend watching the videos of the lectures; they are highly moving: https://www.giffordlectures.org/lectures/discerning-dawn-history-eschatology-and-new-creation.

3

The Origin of Our Species, *Homo sapiens*

Our species, *Homo sapiens*, is believed to have arisen in Africa about 300,000 years ago. The oldest fossil finds, reported from a study published in 2017, are from a site called Jebel Irhoud in Morocco. Except for a slightly thicker brow ridge and a less pronounced chin and forehead, the skeletal remains indicate that the faces of these individuals were very much like our own. The brain case, however, was not. More oblong than ours, it was shaped like a football, quite different than the dome shape of ours. So, although they were not skeletally identical to us, they had enough features in common that most investigators consider them to be an early version of *Homo sapiens.*[1]

Although the oldest specimens found so far are in Morocco, few researchers would assert that our species necessarily originated in this part of Africa. Fossil remains of hominins[2] that we know existed in Africa during this period (200,000 to 500,000 years ago) are exceedingly sparse. The few other sites of this age range that have been identified are geographically spread out—far to the east and to the extreme south. The scientist who led the Morocco expedition, Jean-Jacques Hublin, summarizes the situation in Africa as our species was coming into existence: "What we think is, before 300,000 years ago, there was a dispersal of our species—or at least the most primitive version of our species—throughout Africa."[3] The Sahara was green with lots of lakes and rivers at about this time, and fossil evidence indicates that animals of the eastern Africa savannah were also living in today's

1. See Callaway, "Oldest *Homo sapiens* fossil claim." Also see Stringer and Galway-Witham, "On the Origin of Our Species," 212.

2. Berger et al., "*Homo naledi* and Pleistocene Hominin Evolution." Also see Grün et al., "Dating the Skull from Broken Hill," 372.

3. Callaway, "Oldest *Homo sapiens* fossil claim."

Morocco, suggesting that there was at least some travel back and forth. Today, there is a growing mindset among paleoanthropologists and geneticists that there was not a linear progression within any single African geographical area leading to *Homo sapiens*. Instead, most think there were multiple small populations, each adapted to a particular region of the continent, with occasional mating taking place between otherwise mostly separate, evolving populations. The notion then is that the human body may be somewhat modular, with suites of characteristics that arose in different parts of Africa being combined into an especially versatile and well-adapted combination of traits that came to make up the human body. A key multiauthored review summarizes the basis of this view: "These data are consistent with the view that our species originated and diversified within strongly subdivided (i.e., structured) populations, probably living across Africa, that were connected by sporadic gene flow."[4]

OUT OF AFRICA

Much later—about seventy thousand years ago—members of our species migrated up out of Africa and began their spread through Asia, Europe, and into Australia.[5] There is strong evidence that more than 100,000 years earlier other members of our species had left Africa and settled around the Mediterranean and beyond, but genetic data imply quite convincingly that their numbers dwindled to the point that they did not leave behind offspring whose genes continue to be part of the *Homo sapiens* lineage today.[6]

As the members of our species advanced across Europe and Asia, they came across distant cousins who had descended from other hominins who had migrated up from Africa much, much earlier—almost 2 million years earlier in the case of *Homo erectus*. Members of our own closest cousin species, *Homo neanderthalensis,* had long been present in Europe and western

4. Scerri et al., "Did Our Species Evolve in Subdivided Populations?," 582.

5. There is much discussion about when our species migrated to the American continents. Based on genetic analysis, it is clear that they migrated across the Bering Strait from northeast Asia and there was a land bridge in ancient days. Based on some recent finds in Mexican caves, it appears that it may have been as long as thirty thousand years ago. The migration south may have occurred through a glacier-free corridor that extended down through a region just east of the Rockies, although there is also some evidence that the migration may have been down along the coast of my British Columbia homeland. See Gruhn, "Evidence Grows," 47.

6. There is a remarkably interesting caveat to this statement; a tiny bit of the DNA from that earlier migration out of Africa likely survives in the human genome. See Joseph Akey at the UCSD/CARTA Symposium, February 20, 2020. https://carta.anthropogeny.org/mediaplayer/play/319136/304743.

Asia as well when our species came up. Paleontologists and geneticists think that our species (Sapiens) and the Neanderthals had descended from a single ancestral species, likely in Africa about 600,000 years ago.[7] Based on the fact that some Neanderthal genes are present in all people of non-African descent today, we can conclude there was an interaction between the two species as Sapiens moved into Neanderthal territory. This is not the place to go through all the various lineage studies of the other species already present when Sapiens arrived, but in addition to *H. erectus* and *H. neanderthalensis*, investigators have identified skeletal remains of at least three more species that were present outside of Africa and subsequently went extinct upon the arrival of Sapiens individuals.[8] Strikingly, by 39,000 years ago our species had replaced them all. Adding the extinctions up, this means that Sapiens had replaced at least five species outside of Africa plus at least two more with whom they had previously coexisted in Africa itself. Furthermore, given that the existence of four of these seven species has only been discovered over the past two decades, it is likely that there were more.

So, the question is, what characteristics did our ancestors have that made them so remarkably successful as they moved out of Africa, thriving as they spread across the globe in a manner that had never happened before? Answering this question is significant to us because it is likely there was some suite of characteristics that set our ancestors apart, enabling successive populations to flourish. Ajit Varki, until recently the co-leader of CARTA, a highly influential human origins consortium based at the Salk Institute and the University of California-San Diego, has suggested that a complete geographical "takeover" like our ancestors accomplished may never have been documented in any biological species prior to this. He has called it the "elephant in the room." Just what characteristics did our species have that facilitated such an unprecedented (or at least extremely rare) event? Here is how Varki defined the situation:

> Thus, [*Homo sapiens*] was the "winner" in every contact/replacement event, spanning tens of thousands of years. I cannot find any other example wherein a single (sub)species from one

7. Most investigators have traditionally identified the common ancestral species as *Homo heidelbergensis* based on its anatomical characteristics. However, by no means are they certain about this. More fossil finds will be needed to bring this matter to a more definitive conclusion. See, for example, Grün et al., "Dating the Skull from Broken Hill," 372.

8. This includes a group known as the Denisovans; they are closely related to Neanderthals. They do not currently have a separate species designation, in part because not many of their anatomical details are known. Also, two other species are known to have been present, *Homo florisiensis* and *Homo luzonensis*.

> geographic origin completely replaced all extant cross-fertile (sub)species in every planetary location, with limited introgression* of functional genetic material from replaced taxa,* and leaving no hybrid species. Typically, one instead finds multiple cross-fertile (sub)species, with hybrid zones in between. [* Defined in footnote.][9]

As Varki states, because the original species that previously occupied these world areas had had a great deal of time to adapt to their environment, the expected norm would be for the preexisting cousin-species to continue occupying some of the territory to which they had become adapted, in addition to leaving a hybrid zone bordering the two. The spread of our species such that the native cousin-species were replaced time after time is truly remarkable. Since this phenomenon is exceedingly rare, or maybe even unprecedented in the biological literature, the key question is, what did our species possess that made us so successful relative to the many (at least seven) other *Homo* species, each of which had previously occupied at least one area of the planet along with our species? Varki goes on to address this:

> Although this apparent one-of-a-kind phenomenon could have occurred by chance, the singularity allows one to posit a uniquely complex genetic/biological/cultural transition of [*Homo sapiens*] . . . Adaptive accumulation of an "explosive constellation" of genetic variants could have endowed [*Homo sapiens*] with an unparalleled combination of cognitive features, guaranteeing success at every subsequent encounter with other hominins.* [10]

Similarly, Ian Tattersall and Rob DeSalle, two of the scientists at the American Natural History Museum, have stressed the uniqueness of the displacement:

> Because the pattern previously had been for multiple [hominin] species to coexist, even on limited areas of landscape, this story of rapid replacement everywhere suggests by itself that there was something significantly novel about these early *Homo sapiens* migrants.[11]

9. Varki, "Why Are There No Persisting Hybrids?" "Introgression" is the passing of genes into a species through cross-breeding with another one. "Taxa" is the technical term for a grouping of organisms.

10. Varki, "Why Are There No Persisting Hybrids?" "Hominin" is the term used for individuals of species that are more closely related to the human lineage than the chimpanzee lineage.

11. Tattersall and DeSalle, *Accidental* Homo sapiens, 115.

THE BASIS OF OUR SUCCESS

Until recently, I think most people would have thought that a series of battles from which our species continuously emerged the victor would have been the secret to Sapiens' success as they spread into the new land. This amounts to a sort of "survival of the fittest" scenario, where fittest is defined as the strongest and most powerful. "Fitness" in biology, though, is not defined as the strongest, the fastest, or even the most intelligent. The most "fit" is a technical term related to the number of offspring that live to reproductive age. "Fitness" is, in human terms, a parameter related to family size. For some reason, members of our ancestral *Homo sapiens* species likely had a set of qualities that resulted in greater biological fitness than the other seven or more species they replaced. Varki suggests they were cognitive characteristics, and he has proposed a specific model[12] about this related to the origin of religion that we will address further on page 118. Tattersall and DeSalle also think the key was cognition, while suggesting specifically that it was the acquisition of language and its impact on communication that led to Sapiens' success relative to their less adept cousins.[13] This may well be the case, but Duke University researchers Brian Hare and Vanessa Woods ground the development of our unique cognitive capacity in an unprecedented level of within-group cooperation. Indeed, it is with this in mind that they have entitled their latest book *Survival of the Friendliest.* All hominins in the past half million years likely controlled fire, cooked, ran long distances, and used tools to kill and butcher animals. "This left one important difference between us and all the others: A little more than 50,000 years ago, we experienced a rapid expansion of our social networks."[14]

In his recent book *The Goodness Paradox,* Harvard's Richard Wrangham agrees and describes how cooperation and kindness likely made our species the wunderkind of all hominins that have ever existed. Both books address the problem of renegade activities within groups and between groups, a subject of much theological interest to which we will turn later. But scholars increasingly point to a highly positive group dynamic and cooperativity as the cornerstone to how Sapiens proved to be so successful.

Humans are primates, of course, and primates are known for living in groups with social relationships serving as the cohesive force that maintains stability. The result is a highly successful defensive strategy against predators. Much is known about the neurological networks and strategies that

12. Varki, "Did Human Reality Denial Breach?"

13. Tattersall and DeSalle, *Accidental* Homo sapiens, 122.

14. Hare and Woods, *Survival of the Friendliest*, 63.

play a significant role in group cohesiveness in other primates and, not surprisingly, many of these are important in human cohesiveness as well. However, there *are* significant differences.

JOINT INTENTIONALITY

Like us, chimpanzees have some awareness of the minds of others. They know that others have intentions and can surmise what those intentions may be. This capability likely arose in evolution to aid individuals to more effectively compete for a common food source. Psychologist Michael Tomasello illustrates this through an experiment involving two female chimpanzees who occupy different ranks in their community. The lower ranked female knows her place and is submissive to the more dominant one. In the experiment, the submissive one can see two pieces of food—Piece A and Piece B—but they have been placed in a manner so that the dominant chimpanzee can see only A, not B. The submissive chimpanzee can see her dominant troop-mate spying only Piece A. As they are both given access to the food, the subordinate chimpanzee immediately heads off in the direction of Piece B, ignoring A altogether. Apparently, this lower ranked chimpanzee can understand that her dominant troop-mate had a mind like hers. She understands that they each desire food and that her dominant competitor intends to move to A. So, chimpanzees can understand others as intentional agents, and they are able to mentally simulate what others will do in specific situations. Clearly, chimpanzees possess cognitive skills that enable them to understand, predict, and manipulate not only their physical world, but a sophisticated social dynamic as well.[15]

This "mind reading," however, virtually always occurs in the context of competition and self-interest. Interestingly, chimpanzees do not teach—not even their own offspring. Juveniles learn nut-cracking, for example, by watching others. They receive no help. Tests indicate that chimpanzees are unable to understand when a trainer wants to help them in some way, indicating limits to their "mind-reading" abilities. Chimpanzees will, under certain circumstances, work together and even help one another, but it is only as a means to their own self-centered end. In complete contrast to human children by the age of three—who work together to obtain treats but then know they have a responsibility to share their reward equally—chimpanzees can work together if it is necessary, but after that it is every chimp

15. Tomasello, *Becoming Human*, 12.

for himself or herself when it comes to the bounty: Tomasello indicates chimpanzees have no sense of fairness.[16]

Now contrast chimps and humans. From the earliest stages, babies are uniquely oriented towards others. They have been shown to imitate the facial expression of an adult soon after they are born and engage in proto-conversations with their mother by the time they are just two to three months old. Before the age of one, they usually can understand the intentionality of others. Even babies understand that the other wants to help.

Humans are the only animals that intuitively point to communicate a desire for shared attention.[17] Tomasello states that if you were to go up to any animal in a zoo and point at some object, there are no animals that would understand what that pointing action means.[18] They do not point to draw attention to an object and they do not understand that a pointing gesture is meant to direct their attention to a particular object. Relative to zoo animals, it is exclusively a human trait, and it is a human trait that emerges in the first year of life. Humans understand pointing as a gesture in which one individual assumes that the observer will not only be able to understand what is on the mind of the individual doing the pointing, but wants to share that mindset.[19]

Tomasello and his collaborators have done numerous studies to show that human children are geared towards cooperation from a very early age. They will use a pointing gesture to help an adult find a "missing" object or some other way of communicating if an adult "seems" to be having difficulty in understanding his role in carrying out a cooperative task. As mentioned above, chimpanzees can cooperate if that is what is required for them to get a treat, but each does it only to get what he/she wants. We have already discussed the fact that by the age of nine months, children form dyads

16. Some readers may be aware of some experiments that seemed to indicate otherwise in chimpanzees and capuchin monkeys (e.g., De Waal, *Are We Smart Enough?*, 197). However, Tomasello has pointed to a flaw in the design of those experiments. See Tomasello, *Becoming Human*, 234.

17. Not all cultures point with their index finger. Some point with their lips, nose, or use other ways of drawing an object or situation to the attention of another. See Cooperrider et al., "Protean Pointing Gesture." Also note that some breeds of dogs can be trained to point, but it is not something that they do intuitively.

18. https://www.youtube.com/watch?v=BNbeleWvXyQ. This video of Michael Tomasello's plenary talk at the May 2019, Association for Psychological Sciences annual meeting, is a great summary of the key concepts being discussed here.

19. There is one notable "non-zoo" animal that is an exception to this. From the age of six weeks, dogs can recognize human intent in pointing. The exception, however, is consistent with the rule since dogs have been evolving within human culture for an estimated fifteen thousand years. See this video for a nice illustration: https://youtu.be/8U8zoPQpVbw.

(groups of two individuals) of joint intentionality with a parent. The "I" of an individual essentially becomes a "we." As it matures in toddlers, the "we" is a unit of two persons becoming one in the sense that each recognizes their distinctive role in the partnership, both working in a complementary fashion towards a single common end. As long as the "we" in this partnership have established that "we" are a unit working towards a particular objective, the reward (if there is one) will usually be divided up equally. When one individual does not share, there is generally a sense of guilt for not sharing equally, or at least some sort of acknowledgment of having stepped out of the bounds of their joint intentionality responsibility.[20] Again, I emphasize, chimpanzees seem incapable of acting this way. They might cooperate if that is the only way to get what they want, but once they have what they want, it is each chimp for himself. Observing play among children, on the other hand, one quickly sees a "built-in" sense of joint intentionality—they are a unit. Initially, the joint intentionality is a dyad—just two individuals. However, by the age of three, children develop a sense of collective intentionality: the group size increases to more than two. Again, there are built-in norms; each knows their own responsibility to the group and with that comes a sense of guilt if they deviate from what they sense is best for the collective unit.

THEOLOGICAL CONSIDERATIONS

The theologian Jürgen Moltmann describes the uniqueness of this "we" existence well:

> We see with the eyes of other people. We experience ourselves in the experience of other people. We are in ourselves dependent on the knowledge and recognition of other people. It follows that our experience of ourselves and our empirical awareness of ourselves is always a "mediated" and never an "immediate" self-awareness. Apparently human beings cannot find themselves in themselves, or hold fast to what they are in themselves, without self-division or self-dissolution. It sounds paradoxical, but it is none the less true to say that it is only the person who goes out of himself who comes to himself. It is only in other people that we find the way to ourselves. No one can ever say "I am who I am," for no one is God. We say "I am because you are: you are because I am," for as human beings we are aligned towards other human beings, and as created beings we are aligned towards other created beings. Because from the time we are born, being

20. See the Tomasello video referred to above for a clear demonstration of this.

> human means being-in-relationship, human subjectivity is only possible in inter-subjectivity.[21]

Hans Urs von Balthasar describes the development of this "we" existence and then puts it into a theological context. He starts by describing the role of the mother in drawing out her infant's cognitive awareness of being loved:

> After a mother has smiled at her child for many days and weeks, she finally receives her child's smile in response. She has awakened love in the heart of her child, and as the child awakens to love, it also awakens to knowledge: the initially empty-sense impressions gather meaningfully around the core of the Thou.[22] Knowledge (with its whole complex of intuition and concept) comes into play, because the play of love has already begun beforehand, initiated by the mother, the transcendent.[23]

With that analogy, Von Balthasar then goes on to compare the emergence of the baby's response to her mother's love to how the transcendent God interacts with us as well:

> God interprets himself as love in the same way: he radiates love, which kindles the light of love in the heart of man, and it is precisely this light that allows man to perceive this.[24]

Moltmann emphasizes that the "we" relationship that characterizes who we sense we ought be as human beings mirrors the very essence of who God is: "From one another, with one another, and in one another human beings discover that mirror of the Godhead which is called imago Dei, and which is in actual truth imago Trinitatis."[25] And Pope John Paul II does the same thing when he writes:

> [Humans] became the image of God not only through [their] own humanity, but also through communion as persons . . . the function of the image is that of mirroring the one who is the model, of reproducing its own prototype. [Humans become] an image of God not so much in the moment of solitude as in the

21. Moltmann, *Spirit of Life*, 24.

22. Von Balthasar's use of the term "thou" refers, undoubtedly, to how it is described in Martin Buber's famous book *I and Thou*, where the other person is appreciated as a subject rather than an object; the "I-thou" relationship is one of mutuality and respect, as opposed to an "I-it" relationship.

23. Von Balthasar, *Prayer*, loc. 640.

24. Von Balthasar, *Prayer*, loc. 645.

25. Moltmann, *Spirit of Life*, 221.

> moment of communion . . . essentially the image of an inscrutable divine communion of Persons.[26]

So, the human mind is unique compared to that of other animals in that it is geared toward cooperativity; humans are predisposed to put their minds together and to function as collective units, a concept deeply embedded in Christian theology. How did this "joint" and "collective" intentionality arise in evolutionary history? Do science and theology each contribute to our understanding of this in a coherent manner? It is to these questions we now turn.

THEORY OF MIND

A precondition for this cooperative mindset is the "theory of mind" (ToM) of humans. I like the way that Brian Ware and Venessa Wood summarize what ToM is and its ramifications for the unique qualities of the human mind:

> Pointing is the gateway to reading other people's minds, to what psychologists call "theory of mind." You will spend the rest of your life wondering what other people are thinking. The meaning of a hand brushed against yours in the dark. A raised eyebrow when you walk into a room. It will always be a theory, because you can never really know someone else's mind. Other people have the same abilities you do and can feint, fake, and lie. Theory of mind allows us to engage in the most sophisticated cooperation and communication on the planet. It is crucial to almost every problem you will ever face. It allows you to time-travel and learn from people who lived hundreds and even thousands of years before you. Language is important but fairly useless if you do not know what your audience knows. You can teach only if you can remember what it is like not to know. The political party you vote for, the religion you follow, the sports you play, and every other experience that involves other people, living or dead, real or imagined, all rely on your theory of mind. It is also the soul of your existence. Without it, love would be a cardboard cutout of itself, because what is love without the magic of knowing someone else feels the way you do? Theory of mind is the delight of moments when you both see something, then turn to each other and laugh. It is the comfort of finishing each other's sentences, and the peace in holding hands

26. John Paul II, *Man and Woman*, loc. 3763. Note that John Paul II, writing forty years ago, used the generic term "man." I have substituted this with the word "human."

> and saying nothing at all. Happiness is sweeter if you think the people you love are happy too. Grief is more bearable if you believe someone you lost would be proud of who you are. Theory of mind is also the source of suffering. Hatred burns brighter if you are convinced someone intends you harm. Betrayal is more bitter when you can sift through a hundred memories for every subtle gesture that should have been a warning.[27]

The significance of humans having a full theory of mind and its role in enabling our species to accomplish its countless unprecedented feats in world history cannot be overstated. So, although the statement from Hare and Woods above is a great introduction to the concept, it is important that we define the term carefully. In essence, it entails a recognition that another individual has a mind like one's own. It means that one can have a belief about the intentionality of another.

The minds of animals are sometimes ranked into what is called "orders of intentionality." All conscious organisms are, by definition, aware of their own intentions; they are said to have first-order intentionality. If an organism has a second-order of intentionality, then it can meaningfully construct a belief about what another individual is thinking or feeling, and it can be said to possess a "theory of mind." There are higher levels of intentionality than that, however.

Let's say, for example, that Rick thinks that Jane is under the impression that he is "cool." That is classified as second-order intentionality: Rick thinks he knows what is on Jane's mind and it influences his behavior towards Jane. But we can take this one step further by introducing Bill into the story. Bill knows both Rick and Jane. Bill observes Rick's face when Jane is looking at Rick. Knowing his friend Rick well, Bill sees the look in Rick's eyes, and he thinks he can tell that Rick thinks that Jane is under the impression that he is cool. In this case, Bill is formulating a theory about what Rick is thinking about another mind—Jane's. With that, Bill observes his friend Rick behaving as though he is Mr. Cool, and he knows full well that his friend is putting on an act: he is playing into his perception of Jane's mindset about him. This (Bill's theory about Rick's thinking about what is going on in Jane's mind) is classified as third-order intentionality.

We could keep on going to higher levels of intentionality, but you get the point. Most investigators think that monkeys only have first-order intentionality. Chimpanzees and other great apes, they think, likely approach second-order, but in the view of Oxford anthropologist Robin Dunbar, just

27. Hare and Woods, *Survival of the Friendliest*, 5.

barely. With our much more sophisticated social brain power, most humans, Dunbar states, have between fourth- and sixth-order intentionality.[28]

The process of thinking about what others are thinking is called mentalizing and it is something we do regularly throughout each day. As we write an email, we are thinking about how the other person is going to perceive the message. As we speak, we are thinking about how our statements will affect the minds of others. As we tell stories, we are thinking about the recipient's perspective on what we are saying. We will see below that the process of mentalizing involves different brain regions than do other sorts of logical thinking. The mentalizing network is very highly developed in humans.

Our great ape cousins do have the ability to sense the intentions of the minds of others—but it is rudimentary at best. Dunbar has proposed that one can mathematically relate the sophistication of the ToM capacity of a particular species to the size of the brain region known as the frontal lobe in that species. This is one part of the brain that is especially active when an individual is mentalizing—when a person is orienting their thought processes towards what others may be thinking. In a fascinating graph, Dunbar, based on measurements of fossilized skulls of various hominin species, estimates the size of the frontal lobe in these various species. After calibrating frontal lobe size with the level of intentionality in various living primate species, he estimates (for example) that the species closest to us, Neanderthals, likely had fourth-order intentionality. Here he discusses the ramifications of this observation:

> Mentalizing is crucial for language because both the speaker and listener have to work hard at understanding each other's intentions. The speaker has to intend that the listener understands just what it is that the speaker is trying to persuade the listener to believe, and the listener has to do the same in reverse. The listener might be able to get away with second order intentionality but the speaker probably requires third order and that's before they even start talking about someone else. There is a second reason why mentalizing competences are important for language and that is that the reflexive structure of mentalizing bears an uncanny resemblance to the embeddedness of clauses in the grammatical structure of sentences. The limits on both appear to be at fifth order. Unpacking meaning in language-as-we-know-it can only be achieved by individuals who can cope with fifth order intentionality.

28. Dunbar, *Human Evolution*, 42.

> Early *Homo* populations all sit at third order intentionality, while archaic humans and Neanderthals can just about manage fourth order; only fossil *Homo sapiens* (like their living descendants) achieve fifth order.
>
> What this tells us is that even if the Neanderthals did have language (as many would like to claim), then we can confidently assert that their language definitely wasn't as complex as that which we find in modern humans. Given the crucial role of mentalizing skills in language, the difference in how the two species' brains were organized must inevitably have meant that the quality of the Neanderthal language was very different—in a word, rather more primitive both conceptually and in its syntactic structure.[29]

It is important to remember that although Neanderthals are closely related to our species, we are separated from them by about 600,000 years of genetic changes (along both lineages). Investigators think it likely that the last common ancestral species lived in Africa at that time and at some point, perhaps soon afterward, members of the lineage leading to Neanderthals[30] moved into Europe, leaving behind in Africa those on the lineage leading to Sapiens. The fossil hominin remains from that period are sparse, so it is not possible to trace the step-by-step diversification of the two lineages or even to identify with confidence a particular species as the likely last common ancestor to both. The important point to remember is that Neanderthals were not a precursor to our species; they were on their own evolutionary trajectory until Sapiens came into Europe and displaced them. And, as Dunbar indicates above, and as archaeological remains imply (more about that later), there is good reason to think that their theory of mind was not a full theory of mind, at least not in the sense that characterizes Sapiens.

Evolutionarily speaking, our mind was being developed on a separate and highly diverse continent for hundreds of thousands of years independent of the trajectory of Neanderthals. As discussed earlier, it is likely that our species arose through the occasional mixing of widely dispersed populations throughout the vast continent of Africa. According to this idea, each population would have locally adapted to its own version of the wide variety

29. Dunbar, *Human Evolution*, 242.

30. Although I am referring largely to Neanderthals here, it is important to note that we now know that they have their own cousin group—the Denisovans—which diverged from the Neanderthal lineage about 300,000 years ago. We still have very little fossil information about the Denisovans (although we do have their DNA), so our discussion will focus largely on the Neanderthals as our closest cousin species. In the back of your mind, though, remember that the Denisovans were an equally close cousin species as well.

of environments found in Africa—desert, savannah, jungle, and even the temperate zones of southern Africa. Mixing of genes adapted to a wide variety of habitats might have provided for a rich array of genetic variability that characterized our species as our ancestors migrated up out of Africa seventy thousand or so years ago. This may well be part of the story, but I want to focus on something that I think really gets to the heart of the basis for the remarkable advance in the neurological capabilities of our species.

In a recent study, Robin Dunbar and associates used MRI technology to map and quantify the sections of the brain that are especially active during sophisticated mentalizing (theory of mind) activity in comparison to activities that are just dealing with hardcore facts.[31] Once the brain reaches the sophistication that gives rise to the highest order of intentionality, the investigators conclude, it requires a great deal of brain circuitry. Indeed, much more circuitry is required for mentalizing tasks than for seemingly equally complex tasks dealing only with factual information, without having to take personal interactions and "people-skills" into account.

FULL THEORY OF MIND AND THE DIVINE

So, what came to define our species was an increasingly sophisticated social dynamic and unparalleled language skills—the result of a highly advanced theory of mind with its fifth-order and beyond intentionality skills. How might we integrate theological insights into this overall scientific picture? As the mentalizing capacity of individuals to understand each other in increasingly sophisticated ways approached today's neurological capabilities, how might that have affected the ways in which individuals were able to respond to God's Spirit? They would have been able to think in words now, rather than sensations. Furthermore, they would have been able to think not just with single words, but by stringing those words together into thoughts. What's more, thoughts could be strung together in complex ways that included thoughts within thoughts and rationales about what should and should not be done and why. Consciousness had been taken to a whole new level.

Almost nine hundred years ago, Thomas Aquinas wrote that this unique form of consciousness resulted in our sharing an aspect of the divine nature:

> As far as a likeness of the divine nature is concerned, rational creatures seem somehow to attain a representation . . . imitating

31. Lewis et al., "Higher Order Intentionality Tasks."

> God not only in this—that he is and lives—but especially in this, *that he understands.*[32]

For the first time in the history of life on earth, there was a reflective mind, a change of being, that to some degree—however infinitesimal—had come to resemble the mind of its Creator. Aquinas wrote that "only in rational creatures is there found a likeness of God that counts as an image."[33] It was the expansion of the mentalizing circuitry in the Sapiens' brains that brought it to its current level of sophistication, and most likely, based on the analysis of ancient skull morphology, this had largely taken place by 200,000 years ago. [34]

Anthropologists put great emphasis on the importance of this capacity in shaping our capacity for language, art, religion, and culture in general. It was this mentalizing capacity, they write, that led to the explosive advances that set Sapiens apart—not just from the rest of the animal world—but likely from other hominin species as well.[35] However, in book after book, and journal article after journal article, virtually all scholars examine the emergence of these qualities from the perspective that the natural world is all there is. That, as we saw in chapter 1, is the way science functions—it is called methodological naturalism. However, if there is a God who interacts with humanity through the activity of God's Spirit, this limited perspective misses out on an all-encompassing component in its description of how humanity came to be.

As a child grows and develops an increasing awareness of the minds of others, that child eventually gains a divine awareness as well. As psychologist Justin Barrett has shown, consciousness of the divine matures as the child's discriminatory capacity expands.[36] As Christians, we are confident of the depth of God's love for infants and toddlers; the fact that they are cognitively unaware of that divine love at those stages doesn't make it any less real. However, once the children reach the age at which they understand and, eventually, even "understand that they understand,"[37] children can finally understand the promptings of God, through the Spirit, in a meaningful way.

32. Aquinas, cited in Plantinga, *Where the Conflict Really Lies*, 4.

33. Aquinas, cited in Plantinga, *Where the Conflict Really Lies*, 4.

34. Pearce et al., "New insights."

35. See Dunbar, *Human Evolution*.

36. Barrett, *Born Believers*.

37. This phrase comes from Aristotle, who wrote: "To perceive that we perceive or to understand that we understand, is, for us to be." See Taylor, *Rhetoric, Poetics, of Aristotle*, 357.

At that point, they may or may not respond—the choice is theirs—to the Spirit's urging.

Could it be that there is a parallel between the cognitive development of children today and their concomitant growing discernment of God and what happened in the evolutionary development of our species? God is love; that never changes, but what does change is the ability to understand that love and the call to goodness that flows out from it.

As the Spirit spoke into the everyday lives of the ancient individuals their minds were more socially adept as the centuries and millennia went by. They were able to understand others' intentions better, and Wisdom could be discerned at an increasingly refined level. As the Spirit spoke into their lives, they would have been able to respond in a positive loving fashion (or not), with an increasing awareness of the ramifications of a given course of action. As discussed in the previous chapter, we might expect that not all individuals and communities-as-a-whole would respond positively to God's communication, but let's assume that some would. If so, this would lead to an openness to two-way communication—rudimentary, to be sure, in those days long before Scripture. But God's love for individuals did not begin with the advent of Scripture thousands of years later, and God's loving care would have been continually reaching into human communities, urging individuals to live in harmony with one another and perhaps even with creation in general. Again, I want to emphasize that I am not proposing that all communities would have responded positively to Wisdom's beckoning call, only that some would have; those that did would have thrived and those that did not would have been less successful.

EVOLUTION AND GENETIC CHANGE

Now, I want to bring this back to a bit of evolutionary biology. I start with this statement: If the cohesiveness of a well-functioning community was tightened because members were responding more sensitively to God's presence urging them towards goodness and love, it seems reasonable that such increased cohesiveness would alter the dynamics of the evolutionary process. How so? Forms of genes that fostered group-cohesiveness and the ability to respond to the Wisdom that comes from above would increase in frequency because they would, on average, result in a greater number of children than gene-forms that led to group dysfunction. This is evolution, pure and simple, but it is fostered by the advantage of an ability of communities to respond positively and more wisely to God's presence in their midst.

I need to say a little about the collection of genes that are being selected in this fashion. First it is important to point out that genes typically contribute to building and maintaining the human body as collectives—they work in concert with one another. For example, there are hundreds of genes that influence body height. The genes are collaborative partners; each gene contributes to height-determination in its own way. Each of the hundreds of genes influencing height comes in at least two versions.[38] The genetic component[39] of how tall a person will be is determined by the specific assortment of versions of the hundreds of genes that a person inherits from their parents.

If we similarly consider brain function in the ancient history of our species, there were (and still are) thousands of genes working together to influence the construction and day-to-day functioning of the brain. Of those thousands of genes, which versions of any specific gene an individual had would typically have had only a minor impact on brain function. However, specific assortments would, together, have significantly influenced the probability[40] of certain psychological traits. We could expect that some gene combinations would have led, on average, to a more cooperative disposition. Other combinations would have resulted, on average, in greater mentalizing capacity. Still other assortments of genes would have led, on average, to a more introspective personality.[41] And so on. The point I want to make is that there would have been a variety of personal qualities influenced by genetic makeup that would have constitutionally, on average, primed an individual (and communities of individuals) for wisdom in their interactions with each other and for dispositions more likely to respond to the Spirit's call to live in harmony.[42] Given the likely reproductive benefit (i.e., more children surviving to reproductive age) of thriving communities, combinations of genes that increase the effectiveness of responding to God's call to live lovingly would increase. More cooperation would, on average, lead to more offspring, which in turn would increase the frequency of versions of genes leading to increased harmony and cooperative activity. Over centuries and

38. The official name for a gene-version is allele.

39. Environmental conditions such as nutrition also, of course, influence height.

40. Genes, in general, only influence the probability of a particular outcome; they do not usually determine it. The actual outcome (the official term is "phenotype") generally results from a combination of environmental and genetic factors.

41. In general, genetic makeup does not *determine* the manifestation of a particular trait, it only influences it. Characteristics are influenced by genes, environmental conditions, and—most Christians believe—the transforming work of the Spirit

42. Again, I want to emphasize that genes, in general, do not fix outcome, they only influence it. See, for example, Bliss, *Social by Nature*.

millennia this would lead to an increase in the frequency of the versions of genes that contributed to enabling such dispositions, and a decrease in the frequency of the versions of genes that had the converse effect. In short, this evolutionary step-by-step process took place because there was an ongoing advantage to efficient social interaction. Few human evolutionary biologists would disagree with this statement. What these biologists fail to consider, perhaps because it is outside the domain of science, is that goodness and love have a divine Source, and that individuals in communities responding to the presence of that Source would leave more offspring (on average, over centuries) than those that don't. And this, I maintain, is how our species came to be.

BACK TO THE BIBLE

The challenge, of course, is that the Bible does not directly address this matter. Except perhaps for the story of Adam and Eve, which we will address later, the biblical narrative is squarely placed in the cultures of the ancient Near East. Tools, just four generations after Cain, were made from bronze and iron, we are told in Genesis 4, and Cain himself was the founder of a city. The Bible does not specifically provide us with information about God's interaction with hunter/gatherer communities over hundreds of thousands of years prior to the existence of ancient Near East cultures. However, there are signposts in Scripture that, I think, help to set us on a trajectory that can lead to right thinking about this matter.

For example, Paul, speaking to the pagan Athenian scholars, told them that God had been moving within their culture long before they were fully aware of who God was.[43] Indeed, as Paul introduced them to the God who had been among them, despite their ignorance of the details, he told them it was this God "in [whom] we live and move and have our being." Then, reflecting on what he knew about their culture, Paul added, "As even some of your own poets have said, 'For we too are his offspring.'"[44] Paul told them that they "lived and moved and had their [physical] existence in him," even though they had been ignorant of it, but he went on to say that God would overlook that ignorance no longer. If all of this was true of the Greeks, presumably it was true of the Romans, and if it was true of the Romans, surely

43. "He made all nations to inhabit the whole earth, and he allotted the times of their existence and the boundaries of the places where they would live, so that they would search for God and perhaps grope for him and find him—though indeed he is not far from each one of us" (Acts 17:26b–28).

44. Acts 17:28.

it was true (for example) of the aborigines already living in Australia at the time. And if it were true of them, would it not also have been true of their even more ancient predecessors living in the small communities scattered in different parts of Africa as our species was coming into existence? As they were becoming fully conscious of each other through their increasingly sophisticated mentalizing capabilities, there are tantalizing hints, based on the artifacts that they left behind, that they were also becoming aware of the existence of the Divine.[45] Ought we not consider, like Paul speaking to the Greeks in his day, that even these early members of our species were reaching out to the Divine because the Divine Spirit was reaching out to them?[46] Moltmann speaks of it this way:

> In fellowship with himself and through his creative energies, God the Spirit creates the network of social relationships in which life comes into being, blossoms and becomes fruitful. In this sense "the fellowship of the Holy Spirit" is the activity of the Spirit that confers fellowship or community. Life comes into being out of community, and wherever communities spring up which make life possible and further it, the divine Spirit is efficacious.[47]

THE "GO-BETWEEN" GOD

In his book *The Go-Between God,* John V. Taylor speaks of a walk through the Gallery of the Academy of Florence in which he passes a set of profound sculptures attributed to Michelangelo. Having taken the same walk myself, I concur with his view of the profundity of the experience. However, what is especially interesting is that Taylor puts this walk into the context of the three main characteristics attributable, he thinks, to human features nurtured by the Spirit. First, he describes a set of four sculptures, each one a slave, unfinished, struggling to emerge from the raw stone that still

45. For example, Princeton anthropologist Agustin Fuentes writes: "From the archaeological record we can clearly see that beginning by at least 200,000 to 300,000 years ago, imagination becomes prominent in humans' lives in a way that we often associate with religious experience. The centrality of religious capacity as a feature of humanity solidifies." Referring to this same period, he also states: "Humans began to 'see' and feel and know that their world contains more than the material at hand. They became religious." Fuentes, *Why We Believe,* 145.

46. Note: Paul was only speaking of the Greeks, but his thought on this matter has broader ramifications. I do not mean to imply that Paul had access to any special knowledge about the evolutionary process.

47. Moltmann, *Spirit of Life,* 219.

imprisons them. Just as in those sculptures, the Spirit seeks to draw us out of that which holds us down, helping to shape us into beings free from the constraints of our past. Second, Taylor goes on to one of the greatest sculptural masterpieces of all time: Michelangelo's David. This statue is positioned in the gallery in a manner that allows the observer to circumambulate it again and again, examining David's countenance from all different angles as he was confronted by the mighty giant, Goliath. The viewer is held spellbound at Michelangelo's skill in drawing out the subtle differences in the expressions on David's face. Depending upon angle of view, they are terror, doubt, resolution, and decision. Poignantly, these are also states of mind that each of us may go through during transformative moments when we face life's challenges and respond positively to the presence of God's Spirit providing the wisdom to act rightly in the face of life's challenges. The third sculpture is the unfinished Palestrina Pieta. Here, grief-stricken, Mary holds the limp, lifeless body of her son Jesus, who, as God, had chosen to give fully of himself in a love poured out for humankind. Likewise, the Spirit draws us to love in the manner of Christ even where it means relinquishing our own self-interest. All of this, Taylor tells us, is the way the Spirit of God has been bringing humankind into existence.[48]

Writing almost five decades ago and referring to the work of an eminent biologist of his day, Sir Alister Hardy, Taylor offers this quote from Hardy and then goes on to make an important point of his own from the perspective of a Christian theologian. First the quote from Hardy and a little elaboration of the phenomenon he is discussing:

> I am not in any way belittling the force of the external source of selection . . . but I would suggest that behavioural influence, particularly in the higher group, may often be the more fundamental in determining the animal's make-up . . . It is adaptations which are due to the animal's behaviour, to its restless exploration of its surroundings, to its initiative in seeking new sources of food when its normal supply fails or become scarce through competition, that distinguish the main diverging lines of evolution.[49]

What Hardy describes here is known as the Baldwin effect.[50] It is an established mechanism in evolution whereby the driving force bringing about

48. Taylor, *Go-Between God*, 36.

49. Hardy, *Living Stream*, 132.

50. The Baldwin effect is described especially well on pages 76–77 of *Plausibility of Life*, by Kirchner and Gerhart. It is true that at times single genetic mutations can occur first, followed by behavioral changes now made possible by this genetic change and

genetic change in a population is initially—before any gene changes—a modified *behavior* in response to some circumstance. Perhaps an environmental change would trigger such behavioral alteration, but it could also be just a seemingly random change of cultural *habit* that catches hold. The changed behavior is initially independent of any genetic change; it is merely a change that emerges through inherent flexibility in the behavioral repertoire. To pick an example from within our own lineage, perhaps it involves a way of more effectively making a tool to scrape meat off a bone. Initially, it might be coupled to no new specific genetic feature. But it catches hold in the population and becomes that group's way of doing things. As time goes by, and if the behavior continues to offer substantial benefits to the population, genetic changes might occur that improve the efficiency of this behavior. Perhaps, for example, there is a change in the structure of the hand that makes grasping the tool more efficient or perhaps it is a cognitive change that makes the art of tool construction more effective. The point is that what set the tone for subsequent "hard-wired" genetic change was initially a conscious choice that had nothing to do with genetics and had everything to do with a behavior attuned to a conscious urge to explore new ways of doing things.

We can readily think of other examples of changes that might be initiated in a similar fashion. Consider for example a population that engages in childcare that extends beyond that of the parents to, say, the grandmother. This would be a behavioral change; apes do not do it that way. However, it is not hard to picture the effect that this would have on the reproductive capacity of a population if grandmothers helped their daughters with childcare. Initially perhaps it might be a behavioral change, but there would be even more advantage to a population if a genetic change took place that allowed grandmothers to live longer—beyond their own reproductively active age. So, this is another example of a behavioral change that might establish a cultural change that would subsequently be enhanced by genetic changes.[51]

Taylor then, in writing about the role of the Spirit of God in creation, states:

---

biologists are now in the process of identifying candidate genes. (See Zimmer, "What Makes Your Brain Different," for an example of how a single genetic change—present in our lineage, but not Neanderthals—has likely significantly increased the number of neurons in the frontal lobe of the brain.) However, most behavioral traits are influenced by many genes working in concert. Thus, one would expect that frequently the behavioral alteration occurs first and that later gene-forms accumulate that facilitate this behavior.

51. For a wonderful book length discussion of this point, I highly recommend Hrdy, *Mothers and Others*.

> With the appearance of human reason, choice consciously takes over from chance the direction of the evolutionary process. In Christian terms this means that the Creator Spirit works from the inside of the processes . . . by startling his creatures into awareness and recognition and luring them towards even higher degrees of consciousness and personhood.[52]

Taylor, accordingly, stresses the three main characteristics he envisioned through Michelangelo's sculptures:

1. "[The Spirit] urges every creature again and again to take one more tiny step in the direction of higher consciousness and personhood."
2. "Again and again [the Spirit] creates for every creature occasion for spontaneity and the necessity for choice."
3. "At every turn [the Spirit] opposes self-interest with a contrary principle of sacrifice of existence for the other."

If God is love and if these are indeed three key characteristics of the Spirit of God, it is theologically inconceivable, I think, to imagine the Spirit just leaving those individuals on their own out on the savannahs and other locations in Africa. This would be especially true once our ancestors became fully capable of mentalizing at the level of a "full theory of mind." Once our ancestors had advanced to this stage of neurological development, the promptings of the Spirit would take on a higher, more intense level of meaning and the love that exists only in God would become realizable in a manner never before possible. At this point, individuals could respond in a much more knowledgeable fashion to the Spirit's leading, in loving others and building caring communities.

Paul writes in his Letter to the Romans that those outside of the Jewish tradition in earlier times were accountable to a law written on their hearts:

> When Gentiles, who do not possess the law, do instinctively what the law requires, these, though not having the law, are a law to themselves. They show that what the law requires is written on their hearts, to which their own conscience also bears witness; and their conflicting thoughts will accuse or perhaps excuse them on the day when, according to my gospel, God, through Jesus Christ, will judge the secret thoughts of all.[53]

Who, but the Spirit of God, could have been "speaking" into the hearts of these people? And for how long had this been taking place? Did it

52. Taylor, *Go-Between God*, 33.

53. Rom 2:14–16.

include, for example, as mentioned before, Greeks, Romans, and Australian aborigines equally? Was it restricted to that time period, or had what Paul describes here been going on for many millennia as our species was coming into existence in those African communities?

Although in this age of reductionistic secularism I have not come across recent secular scholars seriously considering the influence of a spiritual force affecting the direction of evolution, this has not always been the case. William James, one of the leading American philosophers and psychologists near the turn of the twentieth century, indicated that he could not accept popular Christianity nor scholarly theism. Nevertheless, he wrote the following in 1902:

> If asked just where the differences in fact which are due to God's existence come in, I should have to say that in general I have no hypothesis to offer beyond what the phenomenon of "prayerful communion," especially when certain kinds of incursion from the subconscious region take part in it, immediately suggests. The appearance is that in this phenomenon something ideal, which in one sense is part of ourselves and in another sense is not ourselves, actually exerts an influence, raises our center of personal energy, and produces regenerative effects unattainable in other ways. If, then, there be a wider world of being than that of our everyday consciousness, if in it there be forces whose effects on us are intermittent, if one facilitating condition of the effects be the openness of the "subliminal" door, we have the elements of a theory to which the phenomena of religious life lend plausibility. I am so impressed by the importance of these phenomena that I adopt the hypothesis which they so naturally suggest. At these places at least, I say, it would seem as though transmundane energies, God, if you will, produced immediate effects within the natural world to which the rest of our experience belongs.[54]

Even though these words are more than a century old, if there is a God, and if that God communicates with us through the Spirit in the way we have described above, they ring true as being in accord with a spiritual influence that shaped the behavior of our ancient ancestors, which in turn could have shaped their genetic makeup through natural selection in the way I have described above (the Baldwin Effect). This is not less true merely because most modern leading evolutionary biologists have long since cast aside the

54. James, *Varieties of Religious Experience*, 270.

possibility of God's existence. As mentioned above, they have done so as part of a worldview choice, not based on a set of tested scientific hypotheses.

About halfway between the time of William James and the present, one could still find the occasional reference to some type of divine presence influencing evolutionary outcomes among highly respected secular scholars. Although Alister Hardy, discussed briefly above, makes it clear that he is not an orthodox Christian, he nonetheless describes an impact of the divine on human evolution not unlike that which I have proposed here. Summing it up, he wrote the following in the mid-1960s:

> It is in this *apparently* non-material part of the world that the power we call God must lie: some source of influence to which Man can have access in an extra-sensory way by the communitive act we call prayer. I have suggested that the power we call God may well have some fundamental link with the process of evolution and saying this I hope I shall not be thought to be belittling the idea of God; I would rather appear to be saying that the living stream of evolution is as much Divine as physical in nature, which is what I believe.[55]

As members of our species became fully conscious of the fact that others had a mind like their own, then what James has written, quoted above, and what Hardy has written in his *The Living Stream* could readily lead to gradual heritable change through the generations. Those with a disposition that resulted in their listening to the Spirit's guidance might live longer lives and be more reproductively fruitful. Those who lived in a manner that was not compatible with smooth functioning within the group and who chose to ignore the inner Spirit-based tugs on their consciences might end up with more stress in life. Their lives might be shorter on average, or they might be partially shunned in some manner that resulted in them being reproductively less fruitful. The main point though is that the God of Christian theology and Scripture is love and based upon what we know about loving connections in families and small groups, it is hard to imagine that God's abiding presence would not be there as they were becoming the species we know today as *Homo sapiens.* Moltmann writes:

> The experiences of God must not be restricted to controlled religious forms of expressing these experiences. We experience God with our whole life, and can use all the ways of expressing life to express the experience of God. If God, the quickening Spirit, is love, then human experiences of love belong within the

55. Hardy, *Living Stream*, 282–83, italics in the original.

> open space of this experience of God; and the experiences of God will intensify the experience of human love.[56]

The early *Homo sapien* understanding of the basis of that presence would have been rudimentary (a bit like Paul describes in speaking of the early Greeks), but that would not deprive them of being fully cognizant of joy that comes from a unity of mind and spirit. Indeed, even today with Scripture to guide us and with Jesus having lived among us for awhile, even we, despite all this, only "see through a glass darkly." Communities and individuals who chose not to nurture, not to teach, not to work together in the hunt for food, and not to care for each other's young would likely not have thrived reproductively. To the extent that the two sorts of personalities—cooperative and noncooperative—were influenced by genetic makeup, one set of genes would become more abundant (say for example, forms of genes that promote friendliness), whereas another set (say forms of genes that promote anti-social behavior) would become less frequent in the species. Responsiveness to the leadings of God's Spirit would shape our physiological, psychological, and cultural attributes, and that—grounded as it is in changes to our genetic makeup—is the basis for the Divine origin of our species.

CONCLUDING THOUGHTS

The Princeton scholar Agustin Fuentes, writing for a secular audience, does not couch it in Spirit-based terms, but he expresses the principle I am grasping at especially well in his inspiring book *The Creative Spark*.

> Recent discoveries and theoretical shifts in evolutionary theory and biology, such as the insights about how our environment and life experiences affect the functioning of our genes and bodies, along with new findings in the fossil record and ancient DNA, have changed the basic story of humanity. A new synthesis demonstrates that humans acquired a distinctive set of neurological, physiological, and social skills that enabled us, starting from the earliest days, to work together and think together in order to purposefully cooperate. Our genes tell only one aspect of how we became creative at increasing levels of complexity. Using these abilities, our ancestors started to help one another care for the young, whether or not those young were their own. They began to share food for both nutritional and social reasons and to coordinate activities beyond what was needed for survival.

56. Moltmann, *Spirit of Life*, 267.

> Acting in ways that benefited the group, not just the individual or family, became increasingly common. This baseline of creative cooperation, the ability to get along, to help one another and have one another's backs, and to think and communicate with one another with increasing prowess, transformed us into the beings that invented the technologies that supported large-scale societies and ultimately nations. This collaborative creativity also drove the development of religious beliefs and ethical systems and our production of masterful artwork.[57]

It is important to emphasize that, scientifically, what I am outlining here is fully within the mainstream of evolutionary biology. As I have already mentioned, most recent anthropologists and biologists have not considered the theological aspects of human evolution, although Fuentes, for one, has indicated that *not* considering the supernatural may be a shortcoming in considering the origin of religions.[58] Most would explain the behavioral changes that set up the advantages of genetic "hard-wiring" as naturalistic cultural adaptations. But this is how science, by definition, works today, and it is why science does not have the tools to address considerations that extend beyond the material world.

One of the great chapters in the Bible is Deuteronomy 30. It consists of Moses's inspiring message to the people of Israel as they prepare to cross the Jordan and enter the promised land. They are a group of people being fashioned into the nation-state into which Jesus will eventually be born. Here is what Moses said to them:

> Surely, this commandment that I am commanding you today is not too hard for you, nor is it too far away. It is not in heaven, that you should say, "Who will go up to heaven for us, and get it for us so that we may hear it and observe it?" Neither is it beyond the sea, that you should say, "Who will cross to the other side of the sea for us, and get it for us so that we may hear it and observe it?" No, the word is very near to you; it is in your mouth and in your heart for you to observe. See, I have set before you today life and prosperity, death and adversity . . . I call heaven and earth to witness against you today that I have set before you life and death, blessings and curses. Choose life so that you and your descendants may live, loving the Lord your God, obeying him, and holding fast to him; for that means life to you and length of days.[59]

57. Fuentes, *Creative Spark*, 5.

58. Fuentes, *Creative Spark*, 208, 216.

59. Deut 30:11–15; 19, 20a.

I love how Moses tells the people about the nearness of God—in their mouth and in their heart. I love the fact that Moses lays out the choice for them: life and prosperity or death and adversity. Choose life, he tells them, so that they and their descendants might live—loving God, obeying God, and holding fast to their God. They were being made into the people of God in preparation for the eventual coming of the kingdom of God, through Christ, for the entire world.

Scores of millennia earlier, God was already fashioning—not a nation (it was too early for that)—but a species, even then with that same end in mind. The eventual goal was no less real, and God was no less involved. Granted, things were different in those African communities while the unique properties of our species were coming into existence. The time span we have been considering is thousands of generations rather than the several dozen for the descendants of Abraham. Furthermore, throughout those many millennia, there was no written law; indeed, there was no writing. But our species was being molded; it was lovingly coming to acquire the genetic and physiological qualities that would set our species apart, enabling individuals to begin to live into the image of God, just as the chosen people of God, the descendants of Abraham and Sarah, millennia later, were set apart, intended to mirror God's image, as they prepared to enter the promised land. Those ancient ancestors were being fashioned into a species that would someday become the people of God and, through Jesus, all humankind would be called to bring the "upside-down kingdom" exemplified by the cross, to all of earth "as it is in heaven." The formation of the biblical nation of Israel was central to God's plan, but we now know that the process began much earlier. Given our theological understanding of the *eternal* nature of the trinitarian God, once our ancestors acquired those socially sophisticated minds that fully recognized the mindfulness of others, God may well have initiated a Spirit-led sense of responsibility to others—just as Moses told the Israelites—that they could either accept or reject. If so, the Spirit of God would have been there offering them the same sort of choices: "life and prosperity" or "death and adversity." This was also the choice that God gave Adam and Eve in the garden—to choose life and prosperity in God's presence or to choose the pathway that leads away from God—one that leads to death and adversity.[60]

In the radio program/podcast *On Being*, the host, Krista Tippett, asked Fuentes what it means to be human. He replied, "The more I studied other animals, particularly other primates and humans, I began to recognize, yes,

60. We discuss Adam and Eve in the context of this scientific knowledge in chapter 7.

humans are animals, but you know, we're a very distinctive animal . . . And so now when you ask me, what is the human? I would argue [a human is] one of the most amazing, challenging, world-changing animals out there with the capacity for *incredible horror* and *amazing love*."[61]

While both capacities are true, of course, nonetheless, in essence it is the "amazing love" part which is most responsible for taking us up out of our animal roots and leading us into a world of music, art, science, and writing. It is the "amazing love" part which has shaped us into the species that put a person on the moon, sequenced the human genome, and is well on the road to bringing diseases like cancer to their knees. There are some fascinating reasons why such "amazing love" is accompanied by our capability for "incredible horror"; we will have more to say about that, too.

In the meantime, and with the assumption that the God of Christian theology is real, we need to learn more about what science has to tell us about how the amazing love of God divinely led to the origin of our species.[62]

61. Tippett, interviewing Fuentes, *On Being*, November 25, 2020.

62. After this had gone to press, I became aware of a book chapter by Jeffrey Schloss which also elaborates on the points made in these last few pages. See Schloss, Jeffrey, P., "Hierarchical Selection and the Evolutionary Emergence of 'Spirit,'" in *The Depth of the Human Person: A Multidisciplinary Approach*, edited by Michael Welker, 125–144. Grand Rapids, MI: Eerdmans, 2014.

4

The Origin of Our Species and the Ability to Imagine

"A LITTLE LOWER THAN GOD"

HUMAN BEINGS ARE ANIMALS too. All aspects of our bodies are structurally, physiologically, and molecularly grounded in our animal roots. Even behaviorally, we can see many of the traits we once thought were specific to humans in other animals. Dogs experience guilt, orcas carry out coordinated planning for a seal attack, dolphins and ravens play, fish learn, and octopi use tools; even fruit flies have culture-specific mating preferences.[1] The better that researchers get at formulating their studies in a manner sensitive to the real-life scenarios that animals face, the more they realize that most of our behavioral qualities are found at some level in the animal world. Kevin Laland puts it like this:

> A long list of strong claims of human uniqueness—humans are the only species to use tools, to teach, to imitate, to use signals to communicate meanings, to possess memories of past events and anticipate the future—have been eroded by science as careful research into animal cognition has revealed unanticipated richness and complexity in the animal kingdom.[2]

Even though animals share many of the qualities we deem special (at least to some degree), that fact does not detract from the poignant passages in Scripture attesting to our uniqueness before God. The following words of

1. See Danchin et al., "Cultural Flies," 1025.

2. Laland, *Darwin's Unfinished Symphony*, 14.

the psalmist of three thousand years ago still ring every bit as true today as they did back then:

> When I look at your heavens, the work of your fingers, the moon, and the stars that you have established; what are human beings that you are mindful of them, mortals that you care for them? Yet you have made them a little lower than God and crowned them with glory and honor.[3]

And the words of Genesis continue to demarcate our unique role relative to creation as a whole:

> So God created humankind in his image, in the image of God he created them; male and female he created them.[4]

The more we learn about the mechanism by which changes in DNA instructions led to the "others-oriented" behavioral shift that stems from a modified brain, the more we realize just how "fearfully and wonderfully made"[5] the human person really is. Archaeological studies make it clear that even though our characteristics are intricately rooted in the world of other animals, with the (divine) origin of our species, these features soar to a whole new level. Beginning more than nine thousand years ago in what is now an Argentinian cave, successive generations of hunter/gatherers prepared a series of wall paintings. Perhaps the most dramatic is a set of hand stencils (see Figure 4.1). Paints were prepared by grinding up and dissolving variously colored earthen pigments, which were then blown out of hollow bone instruments onto the cave wall that had been smeared with an adhesive. We can only imagine what the original artists felt, what they thought, and what they said as they stood back examining their finished project. But that's just the point—they, like us, had the capability of imagining. Chimpanzees and bonobos may on occasion experience the rudiments of imagination; occasionally they have been observed staring out at a majestic scene implying they can appreciate beauty. But they do not *prepare* imaginative art, nor do they stand back contemplating an artist's intentions. It is only in the world of children's cartoons and fantasy movies that we visualize animals working creatively on a project of this sort—mixing and grinding the pigments in solvent, finding the best adhesive material, producing a silhouette of their paw or forelimb, and then reflecting on it. In real life, this just doesn't happen, and we (alone) know it! However, the Argentinian cave is just one

3. Ps 8:3–5.

4. Gen 1:27.

5. Ps 139:14.

of scores of cave walls all over the world known to have been sites where ancient humans set about to prepare stenciled outlines of their hands. The earliest so far—in Indonesia—dates back fifty thousand years.[6] So, members of our species have long been entranced by hand impressions. Although we can be quite certain there are a variety of reasons for this activity, surely one is connected to what the hand signifies to the human mind. It is through the hand, after all, that we frequently reach out and connect with others. Be it a warm handshake, a forlorn departing wave, the pointing action of a baby wanting to draw the attention of her parent to a toy, or the reassuring touch on the shoulder that can serve as a reminder to each of us that our lives matter, the hand facilitates the process by which we open the window into one another's minds.

FIGURE 4.1

Figure 4.1. Cueva de las Manos.[7]

6. Harari, *Sapiens*, 57.

7. Author: Mariano. Creative Commons License.

Hanging on the wall of my own home in San Diego are the two-decade-old impressions of the baby hands of our three grandchildren, Caleb, Noah, and Sara. My wife, Joyce, and I lovingly treasure these little ceramic imprints. They reflect the dimming memory of the beauty we sensed as we touched those tiny baby hands, and the three ceramic molds preserve precious memories of days long gone. By making these hand imprints, even though we did not appreciate it at the time, we were participating in a Sapiens tradition stretching back more than fifty thousand years.

In a book far ahead of its time, *The Phenomenon of Man,* the French Jesuit priest and paleontologist Teilhard de Chardin used a single word to describe human uniqueness. Of all the species in the animal world, we, alone, he wrote, are "reflective." "Because we are reflective, we are not only different but quite other. It is not merely a matter of change of degree, but of change of nature, resulting from a change of state."[8] Those paintings on cave walls have come to symbolize more than just a group of hands reaching outwards from that cold, stone wall. They represent that "changed state," which, on the one hand, brought to life the "amazing love" that Fuentes says represents one of humankind's two most noteworthy characteristics. The other, as he told us, is "incredible horror." Sad to say, the hatred that gives rise to that horror can emanate from that "changed state" as well.

The "reflective" minds of which Teilhard wrote and the capacity to "understand," which Aquinas said is a property of the *imago dei* in each of us, left their prehistorical "imprint" on the world in a way that takes us far beyond handprints on cave walls.

BURIAL SITES

About one hundred miles northeast of Moscow, Russia lies an ancient burial site at a place called Sunghir. There, 34,000 years ago, the bodies of two children about ten and twelve were ceremonially laid out head-to-head, surrounded by various carved objects. One of the children was wearing a fox-tooth hat and a belt that consisted of 250 fox teeth that would have come from at least sixty foxes. About five thousand ivory beads covered each child's body; most likely they were the remnants of a spectacularly

8. de Chardin, *Phenomenon of Man,* 166. The book was largely written in the 1930s, completed in the late 1940s and then finally published after his death in the late 1950s. The book was highly acclaimed by key leaders of the evolutionary synthesis, Julian Huxley (who wrote the Introduction) and Theodosius Dobzhansky (in his own book on human evolution). Similarly, but much more recently, noted evolutionary biologist David Sloan Wilson, in *This View of Life*, xiv, writes that he considers his own book to be largely an updated version of *The Phenomenon of Man.*

decorated coat with this vast array of beads sewn into the now disintegrated covering. Each bead would have taken about forty-five minutes to prepare, so thousands of person-hours were invested in these children's adornment and burial. Alongside the bodies were sixteen carved ivory spears, five of which were double-pointed and the longest of which was eight feet long. The spears were not broken discards; they would have represented the community's best, and they were carefully arranged alongside the children's bodies. Despite their elaborate burial along with these valuable goods, the children showed signs of not having had an easy life. The pattern of enamel on their teeth indicated that malnourishment had, at times, characterized their young lives. One had severely bowed legs and stunted growth. The other had a protruding upper jaw, more extreme—the authors of the study state—than any other skull ever found among Europeans of that era. (See Figure 4.2, which depicts the buried skeletal remains of one of the children.)[9]

FIGURE 4.2

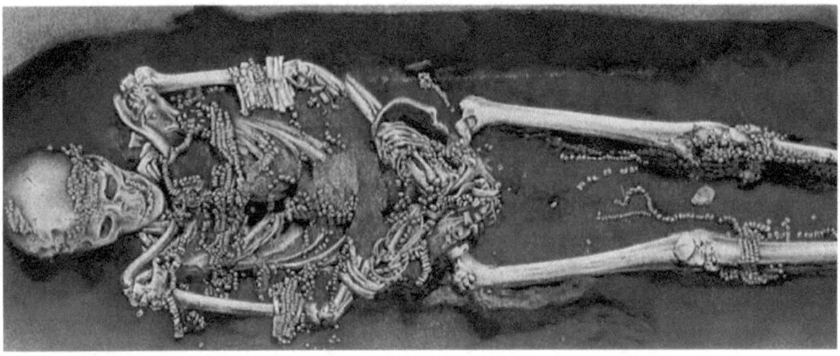

One of Two Children Found Buried at Sunghir[10]

Although we can only speculate why the children were buried in this elaborate manner with the accompanying vast array of labor-intensive accoutrements, evidently their lives mattered in some profoundly significant way to those who were left behind. Too young to have established a special

9. Harari, *Sapiens*, 58. Also, see Trinkaus and Buzhilova,"Diversity and differential disposal."

10. From Trinkaus and Buzhilova, "Diversity and differential disposal." Used by permission.

level of prestige on their own merits, could their physical deformities have been a factor in the amount of attention given to their death and burial? We don't know. However, the willingness of their community members to part with so many valuables in association with the death of these two children implies the existence of a perspective that somehow took them into the perception of a deeper reality, one that transcended their day-to-day experience.

Animals of course do not do anything like this, although some classification groups are certainly aware of the death of others and appear saddened by it. Elephants sometimes cover their dead with soil and vegetation. In one recorded case where an elephant was shot by sport hunters, the hunters returned the next day to find that not only was the body of the dead elephant covered over, but that the large gunshot wound was "patched" with mud.[11]

Like elephants, whales are noted for their high level of cognition. In 2018, the baby of an orca whale named Tahlequah lived for only about a half an hour after birth, but in that half hour the intense mother-child bonding that characterizes the close-knit culture of orcas was established. For the next seventeen days, Tahlequah pushed her dead baby along the ocean surface over a distance of a thousand miles.[12]

Carl Safina describes how a group of chimpanzees embraced each other as they gathered around their dead troop-mate who had just fallen from a high tree. Their facial expression—termed the "fear-face"—revealed the state of their emotions.[13] In another case, a mother chimpanzee was known to carry her dead baby for as long as twenty-seven days before finally letting go.

So, it is likely that the sadness associated with the death of a close companion is part of our animal heritage, but the ceremonial enshrinement of a body no longer alive is found only in our lineage. Indeed, although Neanderthals did sometimes bury their dead, none of the fineries associated with the burials of Sapiens are found at Neanderthal sites. Indeed, an expert on Neanderthals, Rebecca Wragg Sykes, describes a Sapiens burial ground that includes a child laid flat with an "aura" of red ochre pigment and with thousands of small animal teeth and shell beads around her head, shoulders, knees, wrists, and ankles. The ornaments would have taken "months to cut

11. Safina, *Beyond Words*, 68.

12. Safina, *Beyond Words*, 276.

13. Safina, *Beyond Words*, 276.

and grind to size," she writes, as she then goes on to say, "These burials summon up a vision of a child that once laughed and ran as its embroidered clothing shimmered and tinkled. No Neanderthal sites have anything like it . . ." She describes one Neanderthal burial with a baby laid on its side (not flat) with a large red deer jaw. That's it, a red deer jaw. Neanderthals, she states, had an interest in "coloured minerals, shells, and perhaps bird parts, but there's nothing like that with the dead."[14]

In 2000, a large one-mile-long cave was discovered in southwest France. In it were the bones of at least six individuals ceremoniously laid out in several depressions in the cave flooring, some lined with ochre pigment. In an earlier era, the depressions had served as nests for hibernating bears, but beginning 29,000 years ago, they took on a new purpose—they were repository cave depressions for the bones of these six individuals. The ochre lining suggests that the community attached ritual significance to the lives represented by these skeletal remains and likely to other activities taking place in that cave.[15] Especially intriguing is that the remains of the six individuals were relatively complete with one notable body part missing—there were, with one exception, no heads. The community, apparently, was manipulating the dead, although researchers know nothing more than that. Interestingly the walls of the cave are lined with about eight hundred engravings of mammoths, horses, bison, as well as other animals, frequently drawn in poses that depict action. There is a consensus that unlike most caves, this one—the Cussac Cave—was only in use for a generation or so.

The cave contains what researchers call the Great Hall. Characterized by its spacious layout, it has a tall ceiling which fronts onto "the Grand Panel," a thirty-foot-long wall on which many engraved animal figures have been carved, some up to ten feet in size. Intriguingly, many of the drawings on this panel are superimposed on each other, almost like a parchment that was reused from time to time, erasing the old, while overwriting with the new. This, however, was almost thirty thousand years ago and writing had yet to be invented, let alone the use of parchment. But these people did have cave walls to tell their stories and, using their imaginations, they had sharp stone tools to carve story-figures into those cave walls.[16]

14. Sykes, *Kindred*, 337.

15. Kacki et al., "Complex mortuary," 14851.

16. Aujoulat et al., "La grotte ornée de Cussac," 9.

There is no way of knowing the details of what was in fact taking place in that cave, but deep in that underground space with the skeletal remains of individuals they had likely known and loved, the artists of the Cussac Cave were imagining. The sadness experienced by highly cognizant animals like elephants, whales, and chimpanzees in the presence of death had been taken into a new sphere through Sapiens' new gift, that of "reflection." Best of all, they were a community. They were reflecting together.

LION-MAN

One of the most remarkable objects depicting just how far human contemplation had changed from that of other animals is revealed through the scattered pieces of a foot-long statue that had been carved in mammoth-ivory about forty thousand years ago. These fragments were found in a cave six hundred miles away from the Cussac Cave, in southwest Germany. The story of this discovery, the broken remains of the lion-man statue, is a fascinating prequel. On August 25, 1939, two scientists were completing an archaeological dig in a cave in southwestern Germany. The war was about to begin, and they had just received their call-up papers, but that day as they were packing away their tools and equipment to head off to war, they discovered the pieces of a badly fractured one-foot-long statue carved into mammoth ivory, about thirty yards from the cave's entrance. They scooped up the fragments, placed them in a drawer, and went off to battle. It was thirty years later when another scientist, doing an inventory, found the stashed pieces in a drawer. Trying to find matching pieces, he put them together as best as he could, assembling them like one would do in preparing a three-dimensional jigsaw puzzle. Over the next forty years as further digging in the cave revealed more pieces, a total of two hundred pieces were meticulously assembled into the complete sculpture shown in Figure 4.3.

The statue consists of the head and forelimbs of a lion on a slender human (likely male) body. One set of investigators tried to mimic the original carving experience by preparing a replica of the statue using the same sort of stone tools that would have originally been available, and they estimated that it would have taken at least four hundred hours to complete the statue. Given the highly professional artistry, it is quite certain that this person had been preparing ivory carvings for a long time. It is also informative to consider the context of the community that supported this work. This was going on in the middle of an ice age. Nearby, the Alps were buried under nine thousand

FIGURES 4.3A AND 4.3B

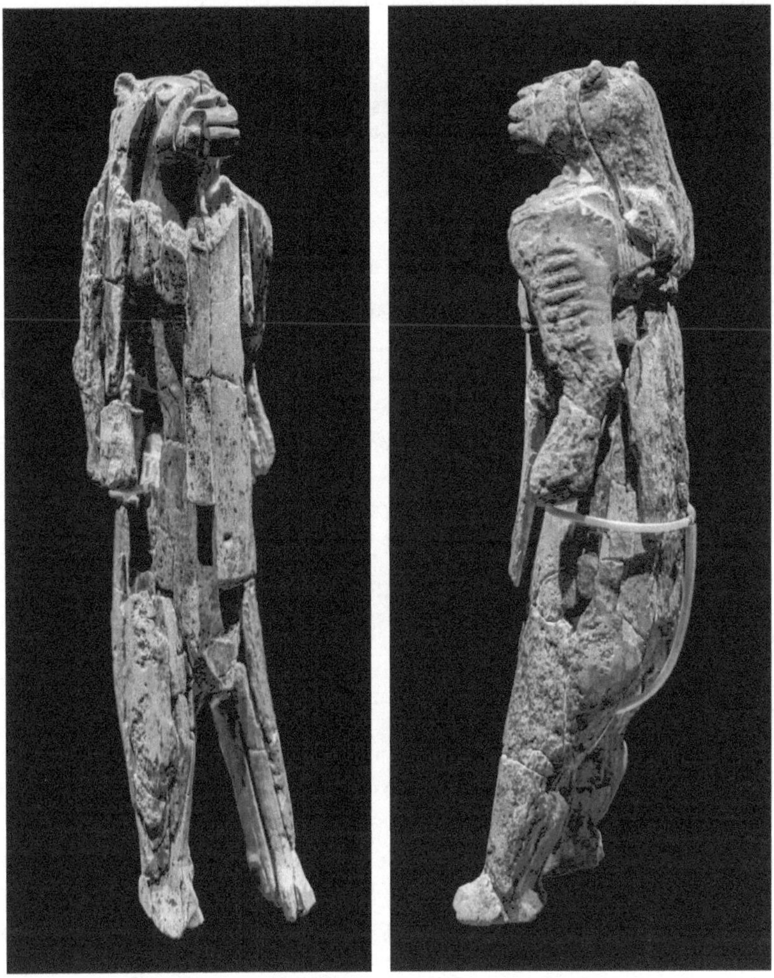

Figure 4.3a and Figure 4.3b Two Views of the Lion-Man.
Sculpted out of mammoth tusk[17]

feet of glacial ice. So even though the spot where the statue was found was in a glacier-free corridor, the community members had to endure long, cold winters; they had a life expectancy of only about thirty years, and they would have been heavily dependent on killing animals for fat and meat to cook on the fire as well as for fur and skins for warm clothing. These people

17. Dagmar Hollmann / Wikimedia Commons; License: Creative Commons Attribution—ShareAlike 4.0.

were eking out a living in difficult circumstances, but still had time for whatever activity was associated with the lion-man sculpture.

This figurine is described in the first episode (October 17, 2017) of a BBC radio program called *Living with the Gods*. The host, Neil MacGregor, is a leading British art historian and as the program proceeds, he becomes engaged in conversation with Jill Cook, curator at the British Museum. "Why would a community, living on the edge of subsistence, whose primary concerns were finding food, keeping that fire going, protecting children from predators, allow someone to spend so much time away from those tasks?" MacGregor asks.

The answer, Cook declares, is that this piece had considerable significance in the life of these people:

> I think this is not just about a relationship to animals. I think it's about a relationship to things unseen, to the vital forces of nature that you perhaps need to propitiate, perhaps to connect to, in order to ensure your successful life. Of course, we know that at this time they were also listening to music. We have a variety of flutes, some made of bird bones, which are already hollow. Some more complex, made from ivory, which again take a long time to make. We have a little figure who is dancing. So, all of these things take us to other places and it's quite possible that this is the case with this figure.[18]

As the conversation between the two continues, MacGregor points out, "The natural irregularities that you expect to find on the surface of mammoth ivory have here been smoothed away—we think as a result of constant handling. This figure seems to have been passed around from person to person over many years—possibly across several generations and what that probably means is stories—not just of the visible world, but of the world or worlds that transcend it to which the lion-man as a transformed being may have given people access."

Cook expands: "I think we must imagine these people sitting around the fire, looking at the magic that the flames create, passing this round, perhaps with stories. But whether this is an avatar to the spirit-world, whether this is an imagining of a creative being of a vital force, we just can't say."

"What is clear," McGregor responds, "is different parts of the world have been put together in an imagined, coherent way."

This was forty thousand years ago; humans had been in Europe for only a few thousand years, and this is about the time that Neanderthals went extinct. Something remarkable had happened to Sapiens.

18. BBC Radio, "Beginnings of Belief."

SOPHISTICATED CAVE ART

I know of no more poignant example illustrating just how far the imagination of our species had come than the Chauvet Cave in southern France. Discovered by three cavers in 1994, it had been closed off by a rockslide for more than twenty thousand years. The cavers found it by hiking along a cliff edge searching for changes in air currents. Upon finding one, they crawled through a tiny space that led into a fantasy world of stalagmites, stalactites, and hundreds of vivid black and red paintings, drawings, and engravings, many in pristine condition. The cavers had no way of knowing just how old the art was at the time of their discovery, but a recent summative study "shows that there were two distinct periods of activity in the cave, one from 37,000 to 33,500 years ago and the other from 31,000 to 28,000 years ago."[19] The figure below shows one example of the faces of four horses. This painting is on a wall adjacent to a whole set of depictions of other animals.

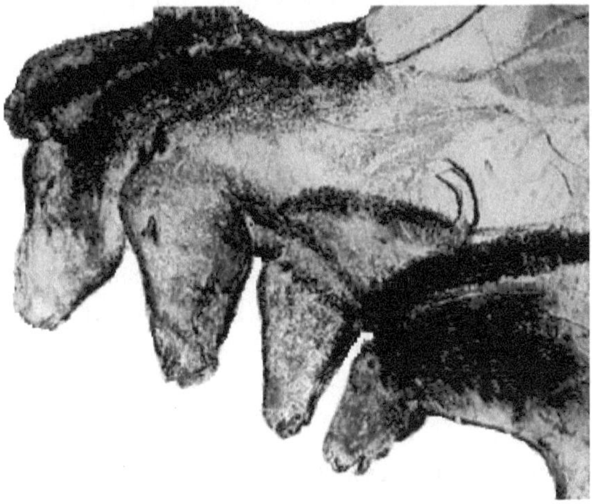

Figure 4.4. Chauvet Horses[20]

In the documentary film, *Cave of Forgotten Dreams*,[21] one of the scientists describes the scene:

> You can also see their keen knowledge of the animal world. They tell a story. Here you have an ensemble of horses, but their open mouths suggest that they are whining. That is to say that these images become audible to us. You see that those two rhinos,

19. Quiles et al., "Chronological Model," 4670.
20. Unknown author, public domain, via Wikimedia Commons.
21. Herzog, dir., *Cave of Forgotten Dreams*.

> there, are fighting. You can see all the signs of fury towards each other. The movement of their legs which are thrown forward and you can almost hear the sound, the thud of their horns colliding against each other in the movement of the fight.

Moving on to another section of wall:

> Here you have another story of a lion, a male, courting a female, who is not ready for mating. She sits and growls. Look, you can hear the female growling—she's raising her lips, she's baring her teeth. She's not happy. And here to finish it off, we have the flight of this bison. You can hear the hooves. We can make out multiple legs indicating its movement. It is escaping from this alcove following this aurochs.

Examining these artworks is an awe-inspiring experience. The artists used the twists and turns in the cave wall to provide a sense of dynamic movement, a three-dimensional experience of a variety of leaping, running, and galloping animals. According to the scientists who have worked in the cave, it is almost as if the animals are alive. Indeed, they say working in the cave is akin to a spiritual experience.

The narrator of the film describes what appears to be an altar at a prominent location in the cave:

> The strongest hint of something spiritual, some religious ceremony in the cave is this bear skull. It has been placed dead center on the rock resembling an altar. The staging seems deliberate. The skull faces the entrance to the cave and around it, fragments of charcoal were found, potentially used as incense. What exactly took place here only the paintings could tell us.

Although we will never know what went on in that cave over the millennia, it is likely that it was highly spiritual in some form or another. As a Christian reflecting on this marvelous scene, I can't help but think back to Paul's words:

> Then Paul stood in front of the Areopagus and said, "Athenians, I see how extremely religious you are in every way. For as I went through the city and looked carefully at the objects of your worship, I found among them an altar with the inscription, 'To an unknown god.' What therefore you worship as unknown, this I proclaim to you."[22]

22. Acts 17:22–23.

One of the other scholars summarized what happened in that cave by saying, "It was not a primitive beginning or a slow evolution, it burst onto the scene like a sudden explosive event. It is as if the modern human soul has awakened here." Another, commenting on the name of our species as *Homo sapiens*, pointed out that "sapiens" means, "the one who knows." Given our ignorance, that is not a good definition at all, he states. Instead, he suggests we should be known as *Homo spiritus* or something like that. By this time, 35,000 to 37,000 years ago, clearly members of our species were reaching out beyond themselves. Given our faith in the God who is love, we can safely assume that the Spirit in turn was reaching out to them—however opaque their understanding of the spiritual reality that they were seeking to take hold of.

Since the oldest paintings at Chauvet cave were about 37,000 years old, this artwork was carried out within just several thousand years after our ancestors had arrived in Europe. This had been Neanderthal territory for likely a couple of hundred thousand years, but they were now gone. Did Neanderthals leave behind any cave art in Europe? The first cave painting attributed to Neanderthals was reported in 2018. It is a rectangle-shaped set of lines on the wall of an Iberian cave dated to about 65,000 years ago, more than twenty thousand years before Sapiens individuals were thought to have arrived in the region. On that basis it was declared to have been done by a Neanderthal.[23] However, in a paper signed by about fifty archaeologists a year later, the Neanderthal origin of even this simple set of markings was called into question and in no uncertain terms. These scientists had two concerns. The first is that it went against "more than one hundred years of research observations on the Neanderthal and modern human archaeological record." But second, and probably more significant, the conclusions were based on a method of dating that can result in considerable misinterpretation if there is leaching of the mineral deposits associated with the painting. As their paper ends, the fifty archaeologist-signatories state: "In short, there is still no convincing archaeological evidence that Neanderthals created Iberian cave art."[24]

As individuals of our species were expanding from Africa up through the Middle East and eventually into Europe, they were also expanding eastward to Indonesia and Australia. The Indonesian island of Sulawesi is especially rich in cave art, dated at more than 35,000 years of age. Indeed, in 2017, an Indonesian caver shimmied up a fig tree to enter a tiny opening, and discovered a particularly remarkable scene, a painting, likely

23. Hoffmann et al., "U-Th dating."

24. White et al., "Still no archaeological evidence."

over 44,000 years of age. The scene depicts wild pigs and jungle buffaloes. Around it, investigators describe small-bodied figures that appear to be human hunters, with spears and ropes, which look like they have animal features, such as a tail and a birdlike beak. The scene is not nearly as well preserved as the Chauvet cave in southern France, but it is of enormous significance. It implies that these people may well have been using art to tell a story. A National Public Radio interview of one of the authors of the study, Adam Brum, puts it this way:

> "They appear to be human, but they seem to have some features or characteristics of animals," Brum says. One appears to have a birdlike head, and another has a tail. He says these part-human, part-animal figures might signal early religious beliefs, because they indicate that ancient humans could imagine things they had never seen. "We can't *know* if it has anything to do with spirituality, but at least we can say that those artists were capable of the sorts of conceptualizations that we need in order to believe in religion, to believe in the existence of the supernatural," he says.[25]

There are two aspects of this story that are especially important. The first is that two widely separated groups of our species were engaged in sophisticated cave artwork at roughly the same time on opposite sides of the Eurasian continent. Neanderthals, even though we have an archaeological record of their activities going back over a span of more than 100,000 years, have left nothing which displays capabilities of this sort. Second, it appears that the artists of the Chauvet Cave, the Cussac Cave, the lion-man statue, and the ornament-crafts persons associated with burials we have discussed in Europe, as well as these artists in the Indonesian caves, show signs of the sorts of conceptualizations that imply that they are aware that there is more to this life than meets the eye. It is likely that they were reaching out to the spiritual realm. The conceptual similarities between what took place at each of the two world areas (Indonesia and Europe), raises the possibility that the activities in which they were engaged began with their common ancestors back in Africa. Given the recency of some of these discoveries (1994, 2000, and 2017), one cannot help but wonder what else remains to be discovered.

The serendipity of these discoveries amazes me. Chauvet was discovered by a slight current of air coming from an almost completely blocked entrance at the edge of a cliff. Cussac Cave went undiscovered and likely was never entered for almost thirty thousand years until just two decades ago, and lion-man was discovered because bits and pieces of ivory were scooped

25. National Public Radio, "44,000-Year-Old Indonesian Cave Painting."

up and placed in a drawer just a few days before the start of World War II. The Sulawesi cave art was discovered because a curious caver noticed a small rooftop opening near the entry to the cave he had been studying. Apparently, he mentally mapped out a route to climb up to that opening, a route which included shimmying up a fig tree. Up he went and upon entering that little hole in the cliffside, he found a sixteen-foot panel with those faded 44,000-year-old paintings.[26] Given the extreme inaccessibility of these sites today (which undoubtedly has been crucial as to why they were largely undisturbed), one must conclude that there was vastly more going on and all we get is a little peek into these elements of our ancestors' life. However, those little glimpses, along with what we know about humankind's inner nature today (see below) give us extremely insightful information about remarkable, unprecedented aspects of human nature and how they came to be.

In most of the cases we have been investigating, there is a clear sense that they thought there is more to life than meets the eye. These people were imagining! They were letting go of the bounds of their day-to-day existence and creatively exploring other "worlds" in their mind's eye.

Regardless how we interpret the details, something remarkable had happened to our species. Neanderthals, it is true, had their own level of sophistication. At a material level, they lived as hunter-gatherers, in a manner that somewhat resembles hunter-gatherer societies of the recent past. Neanderthals controlled fire and cooked their meals of meat that they had acquired through hunting (as opposed to scavenging). Their prey varied from birds to large, dangerous mammals. They likely even participated in rituals. They could make fine blades for tools, and we know they could make pigment, as well as a pitch glue from birch bark that required considerable preparation. However, as Richard Wrangham puts it, "Neanderthals expressed their creativity at a lower rate that *Homo sapiens*. Fewer than ten ornamental beads worked by Neanderthals have been found from 200,000 years of rich archaeology, compared with thousands wrought by *Homo sapiens*"[27] and although that one piece of markings found on the wall of Iberian cave has been attributed to Neanderthals, even that is questioned by many leading archaeologists.

Something specific seemed to have been happening in the generation of our species that dramatically stimulated the ability to imagine, and whatever it was, it apparently had not happened to Neanderthals. As we've seen,

26. To get a feeling for the exploratory aspect of this work and an examination of the cave wall, see this video: https://youtu.be/gx8ohlEAfy4.

27. Wrangham, *Goodness Paradox*, 184.

the two descended from a common ancestral species in Africa dating back about 600,000 years. The lineage that led to Neanderthals likely went up into Europe soon thereafter, whereas those on the Sapiens lineage remained in Africa. Genetic evidence indicates that there was some admixture over time, so there was some wandering, but the lineages remained largely geographically separate. So, the question we addressed in the previous chapter—What happened in the Sapiens lineage?—remains. More specifically, what triggered the astonishing capacity for imagination?

ONE VIEW ABOUT WHAT HAPPENED

In a thought-provoking book,[28] and a follow-up essay in a more recent book,[29] Ajit Varki and his late collaborator, geneticist Danny Brower, have addressed this matter in a novel, intriguing fashion. They point to the enormous advantages of having a full theory of mind—a fifth order of mentalizing skills in Dunbar's terminology[30]—and credit humankind's remarkable triumph to this innovation. Given Sapiens' success story, they ask why this level of cognition has only happened once in the history of living organisms. After all, they point out, other highly cognitive lineages such as elephants, cetaceans (dolphins and whales), corvid birds (the crow family), and chimpanzees have in some cases had tens of millions of years to evolve through that one last cognitive step—to attain their own full theory of mind. Given the tremendous advantage of full cognition (as evidenced by our own evolutionary success), why did it not happen to any of these species, they ask? What could have blocked it? Varki and Brower[31] answer the question by proposing the existence of some huge disadvantage to full cognitive awareness, one that served as a roadblock and prevented further advance along the evolutionary pathway to full cognition. But that just raises one more question: What disadvantage could possibly overwhelm the massive advantages of having an ever more powerful social brain? Varki and Brower propose that the disadvantage stems from a reduction in evolutionary fitness associated with a population becoming fully aware of the ramifications of dying. As they watched others die, full cognition would have enabled the complete awareness of their own death-demise as well. Hardly an advantage

28. Varki and Brower, *Denial*.

29. Varki, "Human Reality Denial?"

30. See chapter 3.

31. Brower passed away before the writing of the book, but Varki attributes the original idea to Brower. The proposal has been fully developed by Ajit Varki, including subsequent talks and writing.

then, this realization would lead to depressing existence. Thus, the reason no other animal reached full theory of mind, they propose, is that whenever a genetic change (or set of changes) took place that advanced their cognition to this most advanced state, any advantages would have been overwhelmed by that one monstrous disadvantage—the depressing realization of their own temporality. As a result, any population of animals—no matter the species—which approached this level of awareness would decline, not increase. The genetic change (or changes) simply could not increase in the population. Whenever individuals became fully aware of the fact that the death of others meant the same will happen to them, it changed the dynamic of staying alive. Thus, the resulting death-anxiety would lead to a decreased evolutionary "fitness," and the more successful evolutionary strategy was to just be kept in the dark about death-eminence. Again, there would have ultimately been a huge advantage to having a full theory of mind, but if the population was not able to make it past that barrier of death-anxiety, the future advantage would be beside the point: it would never be realized. So according to Varki and Brower's model, this block is what prevailed in all other lineages—any genetic changes that led to higher levels of social awareness were just too disadvantageous and they would have died out.

But it did happen to our species! They propose that there is one way in which higher cognition could break through despite the disadvantage. Here is the crux of their thinking: What if a population came into existence where individuals had not only a full theory of mind, but also the capability of imagining some sort of reality that transcended death? Varki and Brower proposed that for each of the two capabilities to take hold—full death-awareness and the ability to imagine death-transcendence—both capacities would need to be "hard-wired" into the brain. Such a population would require two sets of genetically based endowments—one for the full theory of mind and the other for a brain circuitry sophisticated enough to be able to imagine a brand new reality, one that extends life into a realm beyond death. Since the probability of two extremely unlikely sets of genetic events happening within the same population at roughly the same time would be infinitesimally small, that, they propose, is why it has only happened once. None of the several other animal lineages on high level cognitive trajectory (elephants, cetaceans, corvid birds, and chimpanzees, for example), ever crossed the Rubicon to full theory of mind despite millions of years of evolutionary "tinkering." But it happened in our lineage, and that's why they think it did: by chance two independently unlikely set of genetic events happened in the same population at roughly the same time.

I want to emphasize that this idea does not come from the fringes of biology. Varki and Brower's book is endorsed by Nobel Prize winners Peter

Agre and Roger Guillemin, as well as by leaders in evolutionary and cognitive science like Franz de Waal, Simon Baron-Cohen, Francisco Ayala, and Patricia Churchland. Furthermore, Varki himself was, until recently, a leader of what is likely the most noteworthy consortium of human evolutionary scholars in the world.

In a subsequent book, Varki builds upon the theory put forward in the original book. Although denial of death through being able to imagine an afterlife is certainly still an important part of the hypothesis, the emphasis is a little different. Varki now refers to the phenomenon that made the full theory of mind evolutionarily viable as a propensity for "reality-denial," which is defined as follows: "our remarkable ability for ignoring or denying reality in the face of clear facts, a high capacity for self-deception and false beliefs, overarching optimism bias, and irrational risk-taking behavior."[32] According to this hypothesis then, death-denial is one component of a phenomenon that had broad ramifications.

AN ALTERNATIVE PERSPECTIVE

With that as background we now come to the central thesis toward which we have been progressing in this book. Since there is particularly good reason to believe:

- the resurrection of Jesus Christ really took place in history;
- that Christian belief stands up to a vast repertoire of logical scrutiny;[33]
- God's most defining characteristic is love manifest to humankind through the presence of God's Spirit;

then it is only reasonable to consider an alternative to the Varki/Brower interpretation, one grounded not in naturalism, but in a deeper, Christ-centered reality. The reason, we can submit, that humans were able to thrive as they developed a full theory of mind[34] despite becoming aware of their own mortality, was that they were also becoming cognizant of the existence of the eternal. There is good reason, theologically speaking, to believe that God's Spirit would have been at work in their communities, helping them

32. Varki, "Human Reality Denial?,"109

33. There are many books that lay out the reasonableness of the Christian faith. See for example Keller, *Reason for God*, and McLaughlin, *Confronting Christianity*, for example.

34. I want to emphasize that by "full theory of mind" most scholars mean the equivalent of Dunbar's fifth level of intentionality, which we discussed in the previous chapter.

to become conscious of the existence of an additional reality—one over and above the experiences of their everyday world. Once they came to realize this—in response to the work of God's Spirit in their lives—they would have been able to cross the "Varki/Brower Rubicon" and thrive as a population despite each person's cognizance that their own existence on earth is temporary. With that, they would not only have been enabled to deal with death-anxiety effectively, but they would have been able to begin to take advantage of the exciting opportunities made possible by the sophisticated mentalizing capabilities of their minds.

Most secular scientists do not look at the origin of religion in this way. Almost all take some type of "bottom-up" approach, one grounded in the assumption that religion emerged through intrinsic cultural forces alone with no supernatural participation. Recall from our earlier discussions, however, that this focus—exclusively on the natural—is the way science works (by definition). If investigators were to consider input from outside the natural world (that is a "top-down" approach), their study would have moved into the nonscientific realm. To claim, however (as some do), that science leads one to conclude there is an absence of outside influence is false. Given that belief in the existence of the God of Christian theology is logical, it is reasonable to assert that methodological naturalism is only able to tell part of the story. The task of Christian scholars is to build a bridge from what science has discovered so that those findings can be placed within the context of the broader reality revealed through the life, death, and resurrection of Jesus. We are created beings, not the accidental product of blind indifferent forces.

Although it is not possible to determine precisely when members of our species became aware of this grander reality, certainly the sorts of artifacts that we have been discussing—for example, the lion-man of forty thousand years ago and the burial adornment of the Sunghir children 34,000 years ago—imply that their imaginations were beginning to soar beyond the here and now. One thing we do know through anthropological studies is that religion is now regarded as ubiquitous throughout the world and this implies that it emerges from a deep-seated universal that at its core is not limited to certain cultures. Yale psychologist Paul Bloom summarizes the presence of religion in human culture as follows:

> Religious belief and religious practice are human universals. There are no atheist communities and, as far as we know, there never have been. Even within the most secular societies on Earth, the countries of Western Europe, many people are religious to at least some extent, holding certain supernatural

> beliefs (such as life after death) or engaging in certain religious practices (such as prayer). And in the rest of the world—in Asia, Africa, and the Americas, for instance—religious rituals and ideas are at the core of people's day-to-day lives. It is impossible to make sense of most of human existence—including law, morality, war, and culture—without some appreciation of religion and how it works.[35]

Universality of a trait implies that it is thoroughly entrenched into our existence as a species.

Given a "top-down" approach to the origin of our species, there is one principle that must be considered crucial. If there is any one concept that is central to Christianity, it is that God's great desire for humankind is that they live lives characterized by love for one another. First Peter 3:8–9 words it like this:

> Finally, all of you, live in harmony with one another; be sympathetic, love as [siblings], be compassionate and humble. Do not repay evil with evil or insult with insult, but with blessing, because to this you were called so that you may inherit a blessing.

Since the loving character of God has not changed, the prompting to live in that manner has been "whispered in our ear" for as long as we as a species have been able to comprehend it. The choice—to listen, or not to listen—has long been ours, but if that prompting really did occur it would have been a force, working much like natural selection, shaping the formation of our species.

In addition, there are particularly good reasons—grounded in Christian theology—to expect that as humankind became aware of death, that the Spirit of God was, like a gentle breeze, nudging individuals to realize there is more to this world than meets the eye. Indeed, Paul begins his remarkable Letter to the Romans by saying, "Since the creation of the world, God's invisible qualities—his eternal power and divine nature—have been clearly seen."[36] The implication is that once we had the capability of understanding fully, we would have sensed there was an eternal dimension to our lives. Paul goes on in the next chapter to state that even those who know nothing of a written law will sense the requirements of the law "written on their hearts."[37] In other words, based on Scripture, it seems likely that God has always been communicating with humankind about God's existence

35. Bloom, "Religious Belief," 118.

36. Rom 1:20.

37. Rom 2:15.

(however they would have sensed that in those earliest of days) and if that's the case, then as individuals were acquiring those remarkable mentalizing skills they were also developing an awareness, not only of the minds of other individuals, but of the eternal dimension as well.

To recap: Varki and Brower think the evolutionary breakthrough that permitted a full (fifth level of intentionality) theory of mind resulted from two almost simultaneous sets of genetic changes within a single population—one giving rise to a full theory of mind and the other leading to a propensity for "reality denial." Although I have been privileged to spend a little time with Ajit Varki and greatly respect him, I am suggesting that the latter is misnamed. Belief in the nonmaterial is not a denial of reality, not if the nonmaterial really *does* exist.[38] Furthermore, once we consider the "top-down" influence, then the fact that a full theory of mind only developed once, so far as we know, need no longer be considered a puzzle. It happened not because two extremely unlikely events needed to accidentally happen simultaneously. Rather it happened as humans responded to the Divine Spirit. It was because of that presence that humans became aware of an eternal reality. It was this that enabled them to overcome the clear evolutionary disadvantage of knowing they were going to die, and with that, the way to having a full theory of mind was no longer blocked.

I am not proposing that the two (full theory of mind and a growing awareness of the nonmaterial dimension of reality) happened all at once. Moreover, I am not proposing that God would have master-engineered a set of "intelligently designed" genetic changes that would have brought both into existence through manipulating the DNA like a genetic engineer would do. What I am proposing is that as groups of individuals responded positively to the peacefulness of life lived in accordance with the promptings of the Spirit of God (see 1 Peter 3:8–9 above) that the numbers of individuals in such groups would increase. Furthermore, by striving to live into the harmony of life in the Spirit of God, the population would not only become increasingly conscious of God, but they would become increasingly conscious of each other. Genetic changes that enhanced this Spirit-focused and other-centered consciousness would lead to thriving populations, which in turn would increase the frequency of those neurological genetic improvements in the gene pool.[39]

38. Admittedly, humans do tend to deny reality, as Varki profoundly demonstrates in Varki, "Reality Denial." Indeed, this tendency may even be a concomitant side effect of coming to believe in the eternal, but it does not mean that the eternal itself is unreal.

39. It is important to emphasize that this is a logical alternative to the Varki/Brower hypothesis. Varki emphasizes that one of the biggest drawbacks to his hypothesis is that it is not currently testable. What I have done here is to lay out an alternative. If there is a

We know little of how this increased sensitivity to the Divine would have played out through ritual, chants, dance, music, art, and storytelling in those early populations extending over tens of thousands of years. But we can be quite sure that religious expression was manifest in all sorts of ways in the ancient prehistoric past, just as—anthropology has demonstrated—continues to be the case across the planet up to the present. There is no doubt we would cringe at some of the beliefs and practices that existed then just as we do at certain beliefs and practices today. But there would also have been other times when—if we would have been able to peek in—we would have been in awe as we watched the beauty of life being lived in the way we know that human existence was meant to be experienced. It would have been an Eden-like experience and may have functioned so well that it lasted for many generations. All of this is reasonable, and I will have more to say about it in chapter 7.

There is likely nothing more central to living organisms than avoiding death. For hundreds of millions of years success in the animal world has depended upon a survival instinct, and the anatomy, physiology, and behavior of all animals are geared towards keeping them alive long enough to successfully reproduce. We are animals too, and we can be sure that 550 million years of all sorts of genetically programmed death-avoidance is programmed into our DNA, too. Importantly, though, there is a difference between us and animals. Teilhard de Chardin expresses the difference like this: "Admittedly, the animal knows. *But it cannot know that it knows* . . . In consequence [the animal] is denied access to a domain of reality in which we can move freely. We are separated by a chasm—or a threshold—which it cannot cross."[40] That yawning chasm between the other animals and us, de Chardin (like Varki) tells us, is that we are aware of our ensuing death. On the animal side of the chasm, all individuals are unaware that the inner drive to stay alive is ultimately futile. The animal part of who we are battles mightily with the fact that we *have* become aware.

DEATH AWARENESS AND DENIAL

The significance of this awareness on the human psyche cannot be overstated. Ernest Becker, the highly influential psychologist, stated in his Pulitzer Prize-winning 1973 book *The Denial of Death* that the burdensome awareness of each person's ensuing death is the "price that had to be paid," as we

God who interacts with humanity in loving ways, then this alternative is highly logical. However, like the Varki/Brower hypothesis, it is not scientifically testable.

40. De Chardin, *Phenomenon of Man*, 165.

moved into the far reaches opened by a fully conscious existence. Today, we call this the full theory of mind:

> Admittedly, when evolution gave man a self, an inner symbolic world of experience, it split him in two, gave him an added burden. But this burden seems to be the price that had to be paid in order for organisms to attain more life, for the development of the life force on the furthest reach of experience and self-consciousness.[41]

So, the psychologist Becker, like the human-origins scholar Varki, places death-awareness at the center of human existence. Becker did not think there was anything beyond the material realm. And with the perception that modern humans in general were coming to recognize that belief in an afterlife was no longer tenable, Becker stated that to "live on this planet . . . man lives by lying to himself about himself and about his world."[42]

He goes on to say:

> The knowledge of death is reflective and conceptual, and animals are spared it. They live and they disappear with the same thoughtlessness: a few minutes of fear, a few seconds of anguish, and it is over. But to live a whole lifetime with the fate of death haunting one's dreams and even the most sun-filled days—that's something else . . . I believe those who speculate that a full apprehension of man's condition would drive him insane are right, quite literally . . . *Mad* because, as we shall see everything that man does in his symbolic world is an attempt to deny and overcome his grotesque fate. He literally drives himself into a blind obliviousness with social games, psychological tricks, personal preoccupations so far removed from the reality of his situation that they are forms of madness.[43]

Three noted psychologists, Sheldon Solomon, Jeff Greenberg, and Tom Pyszczynski, have addressed this further by putting this into the context of the origin of our species:

> What happened when a life-form, crafted by billions of years of evolution to strive to survive at almost any cost, recognized that it was destined to lose that war? "As a naked fact, that realization is unacceptable," proposed philosopher Susanne Langer, adding, "Nothing, perhaps, is more comprehensible than that people . . .

41. Becker, *Denial of Death,* 263.
42. Becker, *Denial of Death,* 51.
43. Becker, *Denial of Death,* 27.

> would rather reject than accept the idea of death as an inevitable close of their brief earthly careers." Our ancestors consequently used their imagination and ingenuity to stifle their existential dread. They were already employing their sophisticated intellectual abilities to ask and answer questions about how the world worked. But solving the practical problems of living was of little use or solace to them in the face of death. While mountains and stars apparently persisted indefinitely, our ancestors clearly saw that animate entities, subject to forces beyond their control, all came to an end.[44]

The solution to this futility, the authors write, is that "Our ancestors ingeniously conspired to 'Just Say No' to reality by creating a supernatural universe that afforded a sense of control over life and death, enabling them to bound over the 'yawning chasm' and cross the cognitive Rubicon that triggered humankind's evolutionary explosion."[45]

The problem, as I have been trying to emphasize, is that this proposal is a materialistically constrained answer to a question that science alone is not equipped to answer. It simply cannot take its tools into a domain outside of material reality. Indeed, constrained by the limits of its very definition, it cannot explore whether such a domain even exists. With that we turn to the words of Alister McGrath, who said, "Christian theology provides the "theoretical spectacles . . . to rise above the limits of the observable, and move into the richer realm of discerned meaning and value."[46]

One does not have to go back far in time to see that this tendency to overstep the bounds of science was not always as prevalent as it is today. One of the founders of the modern evolutionary synthesis, Julian Huxley, wrote the introduction to de Chardin's *The Phenomenon of Man.* Huxley, in the late 1950s, wrote:

> It is no longer possible to maintain that science and religion must operate in thought-tight compartments or concern separate sectors of life; they are both relevant to the whole of human existence. The religiously-minded can no longer turn their backs upon the natural world, or seek escape from its imperfections in a supernatural world; nor can the materialistically-minded deny importance to spiritual experience and religious feeling.[47]

44. Solomon et al., *Worm at the Core,* 67.
45. Solomon et al., *Worm at the Core,* 64.
46. McGrath, *Darwinism and the Divine,* 289, italics in original.
47. Huxley, in de Chardin, *Phenomenon of Man,* 26.

A second example, also coming from mainstream evolutionary biology is from Dobzhansky, also a founder of the modern evolutionary synthesis. He ended his own book, *Mankind Evolving*, with this quotation from what he called de Chardin's "remarkable" book:

> Man is not the center of the universe as was not evenly believed in the past, but something much more beautiful—man the ascending arrow of the great biological synthesis. Man is the last born, the keenest, the most complex, the most subtle of the successive layers of life. This is nothing less than a fundamental vision. And I shall leave it at that.[48]

Dobzhanky's book was a scientific manifesto about the origin of humankind, and yet Dobzhansky, himself a professing Orthodox Christian, chose to end it with a quotation grounded in the theology of a Jesuit priest! Scientists have very good reasons to turn to Christian thinkers as they seek to understand the bigger-picture ramifications of their work. Ernest Becker, who, as I have already mentioned, was not a believer, nonetheless fills his book, *Denial of Death*, with references that harken back to the faith-embedded thinking of the nineteenth-century Christian philosopher Søren Kierkegaard. Becker's monumental work of the early 1970s is grounded in humankind's search for meaning in light of the "fact" that despite our animal instinct to survive indefinitely, we alone have come to recognize it is not going to happen. As a result, we seek ways to deny death—hence the title of his book. To escape the reality of our ensuing death we seek to suppress it by developing a narrative in which we all become "heroes," in some way or another:

> It is . . . a mythical hero-system in which people serve in order to earn a feeling of primary value, of cosmic specialness, of ultimate usefulness to creation, of unshakable meaning. They earn this feeling by carving out a place in nature, by building an edifice that reflects human value: a temple, a cathedral, a totem pole, a skyscraper, a family that spans three generations. The hope and belief is that the things that man creates in society are of lasting worth and meaning, that they outlive or outshine death and decay, that man and his products count.[49]

In developing his case regarding the significance of death-denial for the human species, he refers to William James, who seventy years earlier had spoken of death as the "worm at the core" of human pretense of happiness.

48. De Chardin, *Phenomenon of Man*, cited in Dobzhansky, *Mankind Evolving*, 348.

49. Becker, *Denial of Death*, 4.

Reminiscent of our earlier discussion of the Varki/Brower hypothesis, James spoke of the significance of religious faith in being able to successfully live well in the face of the human knowledge of ensuing death:

> The lustre of the present hour is always borrowed from the background of possibilities it goes with. Let our common experiences be enveloped in an eternal moral order; let our suffering have an immortal significance; let Heaven smile upon the earth, and deities pay their visits; let faith and hope be the atmosphere which man breathes in;—and his days pass by with zest; they stir with prospects, they thrill with remoter values. Place round them on the contrary the curdling cold and gloom and absence of all permanent meaning which for pure naturalism and the popular science evolutionism of our time are all that is visible ultimately, and the thrill stops short, or turns rather to an anxious trembling.[50]

Writing much later in the century, by which time, he thought, religious belief could no longer be justified as anything but a full-fledged fantasy, Becker's book is largely about how humankind now seeks to come to grips with this "worm at the core" of human existence: death awareness. Interestingly though, Becker goes back—almost longingly, as I see it—to Kierkegaard's victorious solution to death anxiety.

In what follows, I want to look at Kierkegaard through Becker's eyes. It is important that we do so, because the thesis of this book is that our species has come into existence through a God-given assurance that life has a transcendent component, that death does not have the final word. That assurance is grounded in the life, death, and resurrection of Jesus Christ.

Becker tries to come to grips with the ultimate peril associated with dismissing this reality. He thinks that Kierkegaard is profoundly correct about humankind's greatest existential challenge—what Becker terms the human's awareness of being a creature. In Kierkegaard's view there is purpose to the anxiety of death-awareness. Its God-ordained function is to nudge the human toward faith. In continuing to describe Kierkegaard's thinking, Becker writes, "despite one's true insignificance, weakness, death, one's existence has meaning in some ultimate sense because it exists within an eternal infinite scheme of things brought about and maintained to some kind of design by some creative force."[51]

According to Kierkegaard, Becker states, without that leap of faith, there is a sense of helplessness that, upon shedding the facades of

50. James, *Varieties of Religious Experience*, 80.

51. Becker, *Denial of Death*, 90.

character-defenses, leaves one in "sheer terror." William James, even though he was not an orthodox Christian, was saying the same thing. Becker writes that Kierkegaard's Christian perspective enables humankind to break through the cultural "lie" that constitutes a sort of fake purpose to living day after day. Instead, it "opens himself up to infinity, to the possibility of cosmic heroism, the very service of God."[52]

I want to emphasize that Kierkegaard did not conclude that this "opening up to infinity" and its concomitant release from the anxieties of creatureliness, once accomplished, would end anxiety for the rest of one's life. He did not think that faith annihilated anxiety. Quite the opposite. Instead, anxiety remained "ever young" thereby serving as an "eternal spring for growth into new dimensions of thought and trust." Indeed, the anxiety created by the sense of one's own creatureliness continually enhanced faith, creating a lifelong adventure into "multidimensional reality."[53]

So, agreeing with Kierkegaard, Becker maintains that humans were able to find their "heroism," their reason for being in the face of death-anxiety through religious faith. But, writing in 1972, Becker argued, things are now different. Given that it is now recognized as a fantasy, he thinks, religion is no longer a viable elixir for the human condition and the little "heroisms" that used to work to hold the dread of creatureliness at bay are gone. But,

> That is the price modern man pays for the eclipse of the sacred dimension. When he dethroned the ideas of soul and God he was thrown back hopelessly on his own resources, on himself and those few around him. Even lovers and families trap and disillusion us because they are not substitutes for absolute transcendence.[54]

Humankind was now left to its own resources which, according to Becker, now became a matter of largely trying to repress the reality of the anxiety associated with their creatureliness. Humans, he suggests, are destined to try to construct their own form of "heroic activities" that, when push comes to shove, are a sort of reality-denial. In truth, they end up being no more real than the fantasy world of religion itself. The "heroic activities" do not ease death-anxiety; they only repress it, a matter which comes to the forefront in psychoanalysis.

52. Becker, *Denial of Death*, 90.

53. Becker, *Denial of Death*, 91.

54. Becker, *Denial of Death*, 190. Here he is summarizing (and affirming) the views of the psychologist Otto Rank, who had earlier been one of Sigmund Freud's closest colleagues.

I find Becker's conclusions especially relevant to the thesis in this book. Many of our unique human features—the basis of our success—may well have emerged because belief in a transcendental reality made death-awareness tolerable. If this really was the cauldron that gave rise to our species (and the presence of religion in virtually all cultures up until recently supports this), then what happens when the "rug" is pulled out from under our feet? Becker was heavily influenced by Otto Rank, who Sigmund Freud considered to be his heir-apparent prior to a break between the two in the 1920s. Both Rank and Becker considered Christianity to be a profound ideal, albeit one based on fantasy:

> Rank understood that the idea of God has never been a simple reflex of superstitious and selfish fear, as cynics and "realists" have claimed. Instead it is an outgrowth of genuine life-longing, a reaching-out for a plenitude of meaning—as [William] James taught us. It seems that the yielding element in heroic belongingness is inherent in the life force itself, one of the truly sublime mysteries of created life. It seems that the life force reaches naturally even beyond the earth itself, which is one reason why man has always placed God in the heavens.[55]

Considering the solid reasons for believing that Christianity is grounded in a reality that extends beyond the material world, it is not a big jump to declare that the "life force" is not fictional at all. Instead, it is none other than the Spirit of the living God. In fact, Becker, referring to Rank's ideas, goes on to say:

> The "need for a truly religious ideology . . . is inherent in human nature and its fulfillment is basic to any kind of social life." Do Freud and others imagine that surrender to God is masochistic, that to empty oneself is demeaning? Well, answers Rank, it represents on the contrary the furthest reach of the self, the highest idealization man can achieve. It represents the fulfillment of the Agape love-expansion, the achievement of the truly creative type. Only in this way, says Rank, only by surrendering to the bigness of nature on the highest, least-fetishized level, can man conquer death. In other words, the true heroic validation of one's life lies beyond sex, beyond the other, beyond the private religion—all these are makeshifts that pull man down or that hem him in, leaving him torn with ambiguity. Man feels inferior precisely when he lacks "true inner values in the personality," when he is merely a reflex of something next to him and has no

55. Becker, *Denial of Death*, 153.

> steadying inner gyroscope, no centering in himself. And in order to get such centering man has to look beyond the "thou," beyond the consolations of others and of the things of this world. Man is a "theological being," concludes Rank, and not a biological one. In all this it is as though Tillich were speaking and, behind him, Kierkegaard and Augustine; but what makes it uncanny in the present world of science is that these are the conclusions of the life-work of a psychoanalyst, not a theologian.[56]

Informatively, towards the end of the book, Becker continues this line of thought, writing: "obviously all religions fall short of their ideals, and Rank was talking about Christianity not as practiced but as an ideal." Then he goes on to say,

> But as an ideal, Christianity . . . stands high, perhaps even highest in some vital ways, as people like Kierkegaard, Chesterton, the Niebuhrs, and so many others have compellingly argued. The curious thing—as we can now fully appreciate—is that Rank, after a lifetime of work, drew the circle of psychoanalysis itself on this tradition of thought.[57]

What is puzzling to me as I read Becker and his thinking about Otto Rank is that even though each thinks the Christian approach to death-anxiety is grounded in human nature itself, they nonetheless think that it is a fantasy that we must now seek to imitate and replace through psychoanalysis. Given the exceptionally good reasons for believing that Christianity is grounded in reality, why not take the step of recognizing that human nature has come into being because God's Spirit, working through the process of evolution, brought us into being—not as fantasy—but as reality itself?

With all of that in mind, the question remains: Just how did our species change *biologically* to bring about the remarkable suite of psychological and sociological characteristics that became the secret of the success of our species, *Homo sapiens*? It is to that question we now turn.

56. Becker, *Denial of Death,* 174.

57. Becker, *Denial of Death,* 204.

5

The Origin of Our Species and the Cooperative Mindset

INTRODUCTION

We are primates and although our bodies may be like other primates', we alone are the species of "music and art, of poetry and laughter, of science, reason and mathematics."[1] What really distinguishes us is our communal minds. We have devised remarkably successful tools like writing and speaking that enable us to teach and to learn from one another. The usefulness of the tools has extended not just over ever-expanding geographical circles to encompass the entire globe, but it takes us back and forth through time as well. For example, it was the genius of Newton, Maxwell, and many others that fueled the success of Einstein. And it is the words of Paul, penned two thousand years ago in letter form to a small group in Rome, that shape the lives of communities around the world with words like these:

> For I am convinced that neither death, nor life, nor angels, nor rulers, nor things present, nor things to come, nor powers, nor height, nor depth, nor anything else in all creation, will be able to separate us from the love of God in Christ Jesus our Lord.[2]

There is a sense, then, in which human minds have been "wired" together, both "in parallel" (through space) and "in series" (through time). They have been functioning together as a unit and it is the harnessing of that

1. Miller, *Human Instinct*, 4.

2. Rom 8:38–39.

unified power that is the basis of our success as a species. As Yuval Harari has written:

> If you tried to bunch together thousands of chimpanzees into Tiananmen Square, Wall Street, the Vatican or the headquarters of the United Nations, the result would be pandemonium. By contrast, Sapiens regularly gather by the thousands in such places. Together, they create orderly patterns—such as trade networks, mass celebrations and political institutions—that they could never have created in isolation. The real difference between us and chimpanzees is the mythical glue that binds together large numbers of individuals, families, and groups. This glue has made us the masters of creation.[3]

But how did this remarkable mind-power come to be? Here's how Brian Hare and Vanessa Wood phrase the distinction, and we will spend the rest of this chapter discussing its origin.

> All hominins living within the past half million years likely controlled fire, cooked, ran long distances, and used tools to kill and butcher animals. We were not set apart by the size or density of our brain. Other humans, like Neanderthals, had culture within the range of ours . . . This left one important difference between us and all the others: A little more than 50,000 years ago, we experienced a rapid expansion of our social networks . . . The expansion of our social networks initiated a powerful feedback loop. With expanding social ties, we could develop better technology. Better technology allowed us to hunt more food with less effort, which in turn allowed us to feed more people and live in denser groups. Denser groups lead (sic) to even better technology, and so on.[4]

HISTORICAL REFLECTIONS ON HUMAN SOCIALITY

But what explains this rapid expansion of our social networks? Many human evolutionary biologists make note of the fact that there are parallels in the anatomical facial changes that take place when wild animals are domesticated and the change to the smoother, more delicate skeletal features of the face that characterize our bodies compared to the rougher, more robust facial features of ancient ancestral species. They have suggested that the "softening" of facial skeletal features that we know occurs in the domestication

3. Harari, *Sapiens*, 38.

4. Hare and Woods, *Survival of the Friendliest*, 63, 65.

of animals may routinely be a side effect when animals are selected for less aggressive behavior. If this is true, the human skeletal features may be the result of a process associated with the gradual decrease in the aggressiveness of *our* species as well. There are several books[5] and reviews[6] that have been written about this over the past decade or so, and I especially recommend the videos at a symposium for the general public sponsored by CARTA at the Salk Institute in October 2014.[7]

Darwin himself often reflected on this question, but this line of thinking stretches back much further. Aristotle viewed the people in some societies, like the Greeks, in comparison to the hunter-gatherers in the wild, as mild-mannered in a way that resembled horses, cattle, pigs, sheep, goats and dogs. He believed the hunter-gatherers were undomesticated in a manner that paralleled the wild animals from which domesticated animals were derived. In the late 1700s and the first half of the nineteenth century Johann Blumenbach, frequently considered the father of anthropology, wrote extensively about humans in the context of their biological attributes and how they fit into the animal world. Unlike Aristotle, he considered all human populations equally intelligent. "Man," he wrote in 1795, "is far more domesticated and far more advanced than any other animal."[8] In 1806, he went on to write, "The difference between him and other domestic animals is only this, that they are not so completely born to domestication, having been created by nature immediately a domestic animal."[9] As a person who believed in creation, the question of how human domestication had happened was not something he dwelt on. For the farm animals, the mechanism of domestication from their wild counterparts was clear: they had specifically been bred for non-aggressiveness. At one point, within his creationist framework, he briefly mentions that there may have been "a class of higher existences on earth, to whom man acted as a domesticated animal."[10] But the idea was not considered further and, in that age before Darwin, there was no apparent reason to do so.

Darwin, an expert on artificial selection in plant and animal breeding, considered Blumenbach's notion of the human being as a domesticated species, although his focus resembled that of Aristotle. Darwin considered

5. Henrich, *Secret of Our Success;* Wrangham, *Goodness Paradox;* Hare and Woods, *Survival of the Friendliest.*

6. Wrangham, "Evolution of Reduced Reactive Aggression"; Wilkins et al., "Domestication Syndrome."

7. https://carta.anthropogeny.org/events/domestication-and-human-evolution.

8. Cited in Wrangham, *Goodness Paradox,* 51.

9. Wrangham, *Goodness Paradox,* 51.

10. Wrangham, *Goodness Paradox,* 54.

the possibility that only civilized persons were domesticated, but ruled it out for several reasons, including his inability to come up with a potential domesticator. Moreover, he noted, domesticated species were more fertile and with his view of "undomesticated wild men" as being *more* fertile than civilized persons, together with what he considered the lack of reduction in head size (a characteristic of domesticated species), he ruled out the notions of humans having been domesticated in some manner.

Although Darwin was strongly opposed to slavery, he was still a product of his age and his thoughts on human domestication were stymied by the racism of his age. These ideas influenced the direction that his scientific thinking would take him. In essence, he ruled out human domestication because of his inability to fit it into his paradigm, one that viewed domestication as a phenomenon limited to civilized populations. What he did not consider, however, was that the whole species might have undergone a process leading to a lower level of aggressiveness and that his division of humans into "civilized men" and "wild men" was deeply flawed. Darwin is widely considered by biologists (including me) to be one of the greatest minds and one of the greatest experimentalists in the history of biology, but even he and his ideas were a product of the culture of his day.

The idea that *Homo sapiens* as a species was its own domesticator emerged during Darwin's lifetime; Richard Wrangham summarizes it this way:

> For the essayist Walter Bagehot, the contrast of human docility with the aggressiveness of wild animals was too fascinating to ignore. In an 1872 treatise on the theory of political evolution, which Darwin admired and annotated carefully, Bagehot wrote, "Man . . . was obliged to be his own domesticator; he had to tame himself."[11]

THE EMERGENCE OF "FRIENDLINESS"

In this chapter, we will consider the scientific evidence that something like domestication happened in the emergence of our species and that it may have been a significant contributor to our species' success. Although the researchers in this field refer to this as *self*-domestication and most think it was an extremely important component in the evolutionary success of *Homo sapiens,* they are constrained by the way science functions to not consider the possibility that there may be more to this phenomenon than a

11. Wrangham, *Goodness Paradox,* 57.

naturalistic worldview can anticipate.[12] It is important to emphasize, however, that if such a process has taken place in our species, it ought to be considered (at least in part) as a reflection of the factors we've been considering in the previous chapters. Thus, although many scholars think of this as the process by which humankind domesticated *itself*, it is also reasonable to propose that it has happened as a result of humankind's growing ability to sense and respond to the Divine in increasingly sophisticated ways.

The theologian Sarah Lane Ritchie draws her wonderful book *Divine Action and the Human Mind* towards a conclusion with these evocative words:

> This idea—that the mind is a particularly intense locus of divine action because of its remarkable complexity and relational and volitional capabilities—is one worth exploring further.[13]

Although she was not considering the mechanism of evolution in making the above statement, nor was she considering the possibility of humans having undergone a domestication-like event, her work addresses the greater receptivity of the mind to the work of the Spirit as the human mind increased its capacity for relationship and its awareness of the ramifications of its own free choices. If Jesus Christ really did overcome death by coming to life again and the tenets of Christianity really are true (see chapter 1), then it is reasonable to expect that the "intensifying" sense of God's love would have had an escalating effect on the human personality. The parallel between this shaping activity and the human-initiated process of domesticating animals is striking. However, it is a theological concept—one for which science can certainly provide hints—pointing to a horizon that leads into a world of informed faith.

But the key question is, *was* the origin of our species associated with a domestication-like event? If so, how would we know? To tell this fascinating story well, we need to take a little diversion. Our bodies are constructed in much the same way as other mammals, mostly using the same genes, proteins, and even in most cases, entire processes. Indeed, of our approximately 21,300 genes, mice have almost 16,000 of them.[14] A comparison of

12. It is important to point out that although most researchers in this field think that something like domestication occurred in our species, because the process of human evolution is quite distinctive, the question of whether it should be called domestication is not settled. (See Shilton et al., "Human Social Evolution," and Purugganan, "What is Domestication?") Furthermore, even the researchers who stress the similarities between what has happened in the history of our species and the process of animal domestication emphasize that we should expect limitations to the analogy.

13. Ritchie, *Divine Action and the Human Mind*, 346.

14. Yue et al., "Comparative Encyclopedia of DNA Elements."

the 16,000 we share shows that the coding information within each mouse gene is, on average, 85 percent identical to its human counterpart. The processes in other mammals are so similar that in some cases one cannot tell the difference at the early stages between a human embryo and that of an organism that will eventually become very different—like a dolphin or a whale. We each start the embryonic processes according to a single developmental plan, modifying it in species-specific ways as the journey towards an adult-like body proceeds. What this means is that to some extent, what we learn from studying biological processes in one mammalian species frequently applies to humans as well. Keeping this principle in mind, let's look at the domestication process in more detail.

As Agustin Fuentes points out,[15] about 25,000 years ago there were two species of large mammals doing well in the Northern Hemisphere—wolves and humans. They were both highly social, showed a high sense of loyalty to their groups, exhibited communal care for their young, and were excellent hunters. But humans with their big brains, fire, and language had a lot more going for them. Thereby, although there was overlap in the sort of food that they sought, it would have been humans that were the more successful hunters. When humans took down a large mammal like a mammoth, for example, there would be good reason for the wolves to hang around for the spoils. Thus, it is likely that wolves followed human hunting groups. Eventually, there was a subgroup of wolves who became so genetically well-adapted to life alongside humans that they were tame, possibly friendly even, and likely by 15,000 to 20,000 years ago, the domestic dog had become a distinct species. Research indicates that all other mammalian domestication took place after the origin of agriculture and the expansion of villages ten thousand years ago, but dog domestication occurred before that while our ancestors were still hunter-gatherers. However, all those domestication events took place in prehistory and by the time people started writing things down, essentially all our domesticated mammals had already been through the process. There is no clear record documenting the details of the evolutionary change to domestication, with one notable exception. The exception is fascinating.

In 1959, Russian geneticist Dmitry Belyaev began an attempt to tame red foxes. He selectively bred them by choosing those foxes that were least aggressive towards humans as the breeding pairs for the next generation. Year after year, he and his assistants continued this selective breeding process. By the 1970s the foxes in the selected line, in complete contrast to the unselected controls, were wagging their tails in excitement as their

15. Fuentes, *Creative Spark*, 106.

caretakers approached. The videos, readily viewed on You Tube, show the stark contrast of the genetically altered human-loving foxes to their nipping, barking, and highly aggravated cousins. The remarkable point that emerges from these experiments was not only the fact that this personality change occurred so rapidly (ten to fifteen generations), but that a host of other *non*-selected traits emerged with it. Those changes were of the same sort that accompanied domesticated mammals in general. They had floppy ears, a shorter snout, and reduced pigmentation—frequently with white patches on the forehead and other locations. In addition, they had more frequent reproductive cycles and were characterized by more juvenile characteristics both in appearance and behavior. These are all characteristics of domesticated mammals compared to their wild counterparts, and all those changed characteristics (and more) occurred simply by selecting for one trait—a nonaggressive temperament.

Among the physiological changes are the reduced size and function of the adrenal gland. The hormones it produces play a central role in both the fear response and chronic stress level, so a reduced size of this gland results in a more relaxed demeanor. There is also an increased production of the neurotransmitter serotonin in the brain, a change that increases friendliness. So, domestication is associated with a whole set of anatomical and physiological changes, even though the selective regime simply breeds for one trait—nonaggressiveness.

Although not every component of the suite of alterations associated with domestication are found in every domesticated species, many are. Most species have a reduced brain size, reduced skeletal components of the face, and reduced pigmentation. A curly tail is also common. Reduced ear and teeth size are common in many domesticated species—albeit not in foxes. A more prolonged juvenile-like stage is common as are the changes in concentration of some neurotransmitters in the brain, of which serotonin is one.

So, by selecting for changes in one trait—friendliness—the investigators found that a whole lot of other changes took place in parallel. Although we'll discuss one highly respected model for how a somewhat common suite of changes takes place in response to selecting for docility, we need to establish whether something resembling domestication really did occur in our lineage. Since we now know there is a whole suite of skeletal, physiological, behavioral, and even highly specific genetic changes that accompany the "taming" process, it ought not be difficult to characterize those same factors in *Homo sapiens*. One limiting factor, however, is that we do not have the "undomesticated" form to compare it with directly. In dogs, a domesticated species of course, we can compare their characteristics to wolves, the species from which they are derived. In foxes, we can compare the nonaggressive lines to their aggressive counterparts. Even in many of the other domesticated species it is possible to identify species in the wild that are likely closely related to the one that gave rise to the domesticated species we have today. In our own case, however, we do not have that advantage. By almost forty thousand years ago, our ancestors had totally replaced all their cousin species. So how are we going to make the before-and-after comparison?

INCREASED "FRIENDLINESS" AND ANATOMY

Well, for one thing, we do have skeletal remains. Increased sociability is essentially always associated with a modification of the face and skull. Brian Hare and Vanessa Woods described an analysis that carefully characterized changes in the face of many skulls from the ancient past.

> To test our prediction [that domestication had occurred], Steve Churchill and his student Bob Ceiri analyzed the brow ridge projection and face shape of 1,421 skulls, including thirteen Middle Pleistocene skulls from 200,000 to 90,000 years ago and forty-one Late Pleistocene skulls from 38,000 to 10,000 years ago. To measure facial width and height, they used the distance from cheek to cheek and from the top of the nose to the top of the teeth. To measure brow ridges, they examined how far the bone above the eyes projected from the face as well as its height above the eyes. The difference over time was dramatic. The most visible change in the skulls was in the brow ridges. On average, skulls from the Late Pleistocene had a 40 percent reduction in how far their brow ridges projected from the face. Late Pleistocene faces were also 10 percent shorter and 5 percent narrower than the older Middle Pleistocene skulls. While this pattern varied, it continued so that the faces of modern hunter-gatherers and agriculturalists were even more juvenile in appearance than those of their Late Pleistocene ancestors.[16]

So, there has indeed been a dramatic change in the face over time. The alteration in facial morphology is even more dramatic in comparisons which quantitatively analyze the skeletal features of our face relative to that of Neanderthals.[17] The "softening" of the skeletal features of the face, which seems to have accompanied the origin of our species, has been a telltale

16. Hare and Woods, *Survival of the Friendliest*, 69.

17. For an analysis of the technical details in this sort of analysis, see Hublin et al., "New fossils from Jebel Irhoud." For a less technical discussion, see this video: https://carta.anthropogeny.org/mediaplayer/play/319138/304736.

sign of a tendency towards tamer personalities in animals that have been domesticated. Reduction in the skull size is another, and indeed the average size of our skull, at 1,350 cc, is 10 percent reduced from the 1,500 cc average that characterizes Neanderthals.

There is another skeletal signature, besides the face and size of the skull, that serves as an indication for a lower level of aggressiveness. Androgens are a class of hormones that are associated with increased aggressiveness. In mammals, if a mother pumps a higher concentration of androgens into her developing embryo, the digits of the front limb are formed a little differently. For reasons that need not concern us here, the equivalent of the index finger does not grow as long as the equivalent of the ring finger. This is called the 2D:4D ratio (the second digit to fourth digit ratio). This ratio—a reflection of events in the embryo—is correlated with species' aggressiveness. The correlation between this embryonic process and temperament has been demonstrated in a variety of species, but one especially poignant example is a comparison of bonobos to their cousin species, chimpanzees. The two species have evolved separately for an estimated 900,000 years and are noted for their vast difference in personality. Chimpanzees can be extremely aggressive; bonobos are much calmer, readily settling their differences through grooming and sexual activity. Sure enough, the index finger of the chimpanzee is shorter than the fourth digit, whereas in the bonobo this difference is not observed.[18]

So, with that as background, we can ask whether the difference in length between the two digits has been changing in our lineage by analyzing fossils. The answer is yes, the 2D:4D ratio has indeed changed in a manner that parallels the gracilization (smoothing out) of the face over time, as discussed above. This is compatible with the development of an increasingly friendly demeanor over the past 100,000 years. And, not surprisingly then, measurements show that Neanderthals have an even lower 2D:4D ratio compared to Sapiens.[19]

Thus, there is reason to suspect that something analogous to domestication took place in our species. But we need to be careful about jumping to conclusions based upon changes in skeletal features alone. It is important to

18. This is not the only feature which signals that bonobos have a lower level of aggressiveness. They also have a higher concentration of serotonin-releasing neurons in the amygdala region of their brains. Based on many studies, investigators think this would likely result in a calmer demeanor. Bonobos have a less prominent brow ridge, their facial bones are not as pronounced, their jaw does not protrude as much, and they have a less developed muzzle. These are all characteristics of lower levels of aggression and general features of the tameness of domesticated animals.

19. Hare and Woods, *Survival of the Friendliest*, 50.

address whether a broader suite of characters have been altered in humans in a manner that parallels the domesticated animals. To answer that question, we need to explore what it is about the mammalian body that results in this suite of changes that accompanies increased sociability.

INCREASED "FRIENDLINESS" AND EMBRYO FORMATION

Recall from our earlier discussion that the embryos of all mammals begin the developmental process in much the same way. The leading model to explain the basis of domestication in animals is a tweak to an embryonic process that all vertebrate[20] animals share. If this process is altered, many investigators believe, it brings about a somewhat uniform set of changes, one of which is reduced aggressiveness.

So, what is the process and how is it altered when an animal becomes less aggressive? The answer is fascinating. As the embryo is being formed there is a tube that runs along its back and up into the region that will become its head. This becomes the spinal cord and, at the head end, the brain. There are some cells with a special property that are found just above this tube in all vertebrate embryos; they are called neural crest cells. They migrate out from where they are first formed and, after multiplying, start developing into various components of the body. Some become part of the face and the skull. Others become teeth. Some develop into pigment-producing cells. Others develop into parts of the peripheral nervous system and, especially noteworthy, the adrenal gland. This is summarized in the figure below. Note that in this diagram, the neural crest and the tube that will become the spinal cord are shown in an adult body. This is so that you will not be distracted or confused by the technicality of the embryonic body form. The important point on which to focus is the spreading out of neural crest cells from their initial location at one edge of the tube that will become the spinal cord.

20. Vertebrate animals include all animals with a backbone; hence mammals are vertebrates. We are focusing here on mammals, but other vertebrates, like chickens, have been domesticated, and they have a set of other characteristics associated with their nonaggressive personality that fit into the domestication suite.

FIGURE 5.1

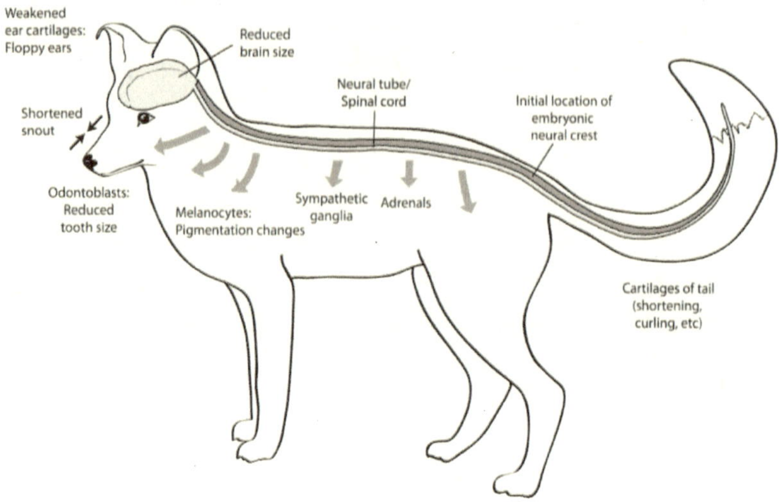

Figure 5.1. Movement of Neural Crest Cells. ***Note the neural tube/spinal cord running along the back of the body and, just above it, is the location from which neural crest cells will migrate. As they migrate, they mature into the cells that make the teeth, bone of the face, pigment producing cells, adrenal gland, cartilage of ears, and tail. They do not migrate into the brain to produce parts of the brain, but they do influence brain maturation indirectly.***[21]

In 2014, three investigators proposed that the basis of domestication was a set of genetic changes that affect the migration of the neural crest cells.[22] According to the hypothesis there is a reduced number of these migrating cells arriving at their appropriate destinations to fully contribute to their designated bodily structures. Hence, the teeth are smaller, the face is less pronounced, and the snout does not extend out as far. Even the pigment producing cells do not always make it to their intended locations which results in white patches—especially in extremities. The ears are often floppy because there are not as many neural crest cells developing into cartilage at the outer edge of the head where ears are formed.

What about the all-important behavioral changes? The answer, in part, relates to the reduced size of the adrenal gland. The adrenal gland produces hormones that can cause high levels of reactivity and stress. Smaller adrenal

21. From Wilkins et al., "Domestication Syndrome." Used by permission.

22. Wilkins et al., "Domestication Syndrome."

glands mean less reactionary fear and stress and, thereby, a tamer demeanor. The brain itself is also altered in domesticated animals in ways that are consistent with a calmer and more social disposition. The effect on the brain is noteworthy because the neural crest cells do not directly contribute to brain tissue. However, they do have an indirect effect on brain development, so fewer neural crest cells would mean a weaker indirect influence. Hence, the reduced size of the brain and the changes that foster a less aggressive, more social temperament.

So, according to this model, the suite of changes associated with domestication is due to genetic changes in neural crest cell formation and/or migration. In the fox experiment, when the breeders selected for tamer foxes, they were choosing those animals that developed from embryos with slower moving and/or a reduced number of neural crest cells. Just like what happens when a dog breeder selects for certain desirable traits, when foxes are bred for tameness, the process gradually enriches for genes that slow down the formation or migration of neural crest cells in general. As a result, not only are the foxes tamer, but they also have shortened faces, more white patches, floppy ears, curly tails, and altered hormone production because the genetic variants that cause tameness do so through their negative effect on the formation and/or migration of neural crest cells in general.

That is the model, and it makes good sense, but it is so central to who we are as a species, I want to address the fascinating experiments that imply rather strongly that the model is on the right track. To do that, we need to spend a little time getting you, the reader, brushed up on a few genetic concepts from your high school biology class. Following that we will get back to a discussion of how recent data suggest that altered neural crest cell migration may have been one of the important components in bringing about our species' unprecedented degree of sociality.

INCREASED "FRIENDLINESS" AND GENETIC CHANGE

Gregor Mendel was an Austrian monk that you would likely have heard about in high school biology. He is considered the "father" of the science of genetics. When I was a genetics professor, I used to begin each course by refreshing students' memory of the rules of inheritance that Mendel discovered by studying abnormal strains of pea plants. For example, one abnormality resulted in the seeds being wrinkled rather than smooth. Clearly, seed texture was influenced by a gene that existed in at least two varieties; one caused seeds to be smooth, its alternative resulted in wrinkled seeds. Can we identify something like that in humans as it relates to friendliness or nonaggressiveness? If indeed we became more friendly during the origin of our species, and if there was a genetic basis for that change, might we be able to follow the inheritance of such genes in a manner that parallels the inheritance of seed texture in Mendel's peas?

The story that emerges as we examine this is intriguing. It begins with the discussion of a genetic anomaly called Williams syndrome.[23] Persons with Williams syndrome have a very cheerful disposition and they are super-friendly. Born at a frequency of about one in ten thousand, the children with this genetic abnormality quickly endear themselves to others, a quality that continues into adulthood. They are hyper-social and exhibit a spirit of happiness. Unfortunately, though, they also have some physical problems and abstract learning is usually difficult for them. Their IQ is in the fifty to sixty range, and they often have cardiovascular issues that require multiple surgical interventions. Kidney dysfunction is another challenge associated with Williams syndrome. Although their general IQ is low, there are some activities at which their abilities really shine. For example, they frequently have well above average musical interest and skills, which are associated with a high level of sensitivity to sound in general. They are characterized by reduced tooth size and the lower half of their face is associated with distinctive features. All these changes are due to a highly specific genetic alteration, which I describe below.

To begin, I need to introduce a few fundamentals of genetics to make sure we are all on the same page as we seek to understand the basis of the distinctive characteristics of individuals with Williams syndrome. I will also restate a few ideas that I discussed in earlier chapters to refresh your memory. Humans, as we have discussed already, have about 21,300 genes. They are distributed on twenty-two chromosomes plus one set of distinctive chromosomes, the X and the Y chromosome, which are important in determining gender. Each distinctive chromosome (except the X and Y) is given a number between one and twenty-two ranked by length where one is the longest. The average chromosome has about a thousand genes.

Each chromosome contains one long, thin thread, a single DNA molecule, stretching from one end to the other. Each of the thousand (on average) genes is a segment of the molecule and each usually contains information on how to produce a protein or enzyme to carry out a particular function. Usually, that function is simply providing the instructions for making a particular protein molecule within specific cells of the body. The DNA instructions for constructing that protein are copied much like an

23. The Williams Syndrome Association has some fine videos explaining the characteristics of the syndrome and includes interviews with some with the syndrome. See for example: https://williams-syndrome.org/what-is-williams-syndrome.

engineer on a building project might photocopy the blueprint plans for that project before passing out copies to everyone to see.

Genes influence our traits through the protein-building instructions they provide. Some abnormal traits are caused by an alteration in a single gene. For example, cystic fibrosis is caused by an abnormality in a gene that specifies the structure for one of the proteins needed for normal lung function. Were it not for medical treatment, this abnormality would result in death prior to adulthood. So cystic fibrosis is a trait grounded in altered coding instructions of a single gene. Often, however, traits are *multigenic*; they result from the actions of many genes working in concert. Height, as mentioned in chapter 3, is heavily influenced by many genes.

There is one other consideration that is foundational to our discussion of some genetic factors that influence sociability. Although twenty-three chromosomes have the complete set of genes needed to build and maintain the human body, we in fact carry forty-six chromosomes (two of each of the twenty-two non-sex chromosomes, plus two sex chromosomes; males have an X and Y chromosome, whereas females have two X chromosomes), meaning that most genes are represented twice. One of the copies has come from our mother, the other from our father. Thus, we essentially have two versions of most of our genes, referred to as *alleles*. For any given gene, the maternal version may be a little different than the paternal.

The processes involved in making and maintaining our bodies are tightly attuned to having two copies of most of our genes. Although the body can often tolerate changing this balance for a single gene or a tiny group of genes, in general it is extremely important that there be two copies—not one, not three, but two. Sometimes deletions or duplications of a chromosome segment occur and are passed on to the next generation through the sperm of the father or the egg of the mother. When that happens, it breaks the two-copy rule and almost always results in lethality of the embryo, unless the deleted or duplicated segment is exceedingly small.

Williams syndrome is a case in point. These individuals carry a tiny deletion of less than 1 percent of Chromosome 7. Since their other copy of Chromosome 7 is normal, that means they have one, but only one, copy of the twenty-eight genes that the deletion has removed. Having just one copy of this block of twenty-eight genes means that there is generally only half as much of the protein products that each of the twenty-eight genes specify. This shortage of protein products creates abnormalities that make life difficult for these individuals, but as mentioned above, they are generally friendly and thrive on being with others.

If having just one copy of this tiny set of genes results in an excessively social disposition, it would be interesting to know what would happen if a person had three copies. A chromosome with a duplicate copy of the same twenty-eight genes missing in Williams syndrome does come into existence from time to time. Interestingly, individuals who carry this chromosome with its two copies and a normal chromosome with its single copy (three in all) have the opposite personality to those with Williams syndrome. Usually they are socially withdrawn and, very frequently, are on the autism spectrum.

Since a gene contains the plans for making a protein molecule, what this tells us then is that Williams syndrome (with its half the normal number of these genes) is the result of having only about 50 percent of the protein products of one of those genes in the deleted region of Chromosome 7. The alternative opposite syndrome is a result of having about 150 percent of the product of that same gene. That gene has been identified, and for our purposes of keeping things nontechnical I will call this Gene 1.[24] Interestingly, this same gene is also associated with the degree of sociability in dogs, where it likely played a role in their domestication from wolves.[25] Indeed, even in mice, experiments have shown that Gene 1 influences sociability and does so through a specific change in the brain.[26] Gene 1 is known to be a master gene that controls the reading of other genes, some of which produce proteins important for brain function (see Figure 5.2). The fact that this gene likely played a role in domestication in dogs and that it can influence sociality in mice implies that it could have been a target in affecting the level of sociality in humans. Presumably tweaking how frequently the instructions in this gene are read could have had a major impact on the sociality of our species compared to sister hominin species, but demonstrating that was the case is no simple matter and as of this writing, the key experiments have not yet been done.

24. The official name of what I am calling Gene 1 is GTF2I. For evidence that this is the key gene see Tassabehji et al., "GTF2IRD1 in Craniofacial Development."

25. BvonHoldt et al., "Structural Variants."

26. The effect of this genetic change on the brain is a shift in the balance between certain neurotransmitters. Li et al., "Induced Chromosome Deletions."

FIGURE 5.2

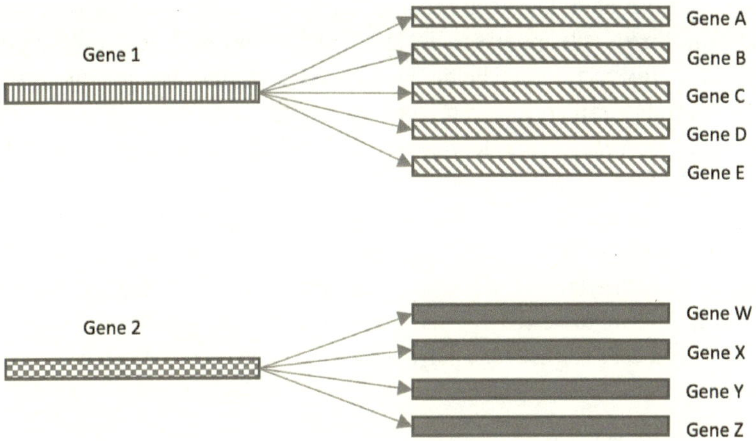

Figure 5.2. Gene 1 and Gene 2 are Master Genes. ***They each produce a transcription factor that controls the level of expression of other genes. This is an idealized diagram to illustrate the concept.***

However, there is a second gene of the twenty-eight genes deleted in Williams syndrome for which comparisons with Neanderthals have recently been completed. I will describe the work on this gene briefly, partly because the results demonstrate that this gene, which is also important for sociality, has undergone changes in our lineage in ways that likely would have increased the sociality of our species.

Recall that our face has been significantly altered over the last several hundred thousand years; it became "softer"—increasingly gracile is another way of putting it. Interestingly, persons with Williams syndrome are noted as having elf-like facial features. It appears that the basis of this increased facial gracility for Williams syndrome persons can be narrowed down to just one of those twenty-eight missing genes in that deletion. For the sake of readability, we will call this one Gene 2 even though like Gene 1 it does have an official name.[27] Some elegant experiments demonstrate that Gene 2 likely functions in a manner that regulates the amount of neural crest cell

27. The official name for this gene is BAZ1B. The scientific paper which describes the molecular details I outline here is Zanella et al., "Dosage Analysis of the 7q11.23." It is summarized at the general level in Price, "Early humans domesticated themselves."

migration, which thereby affects the formation of the skeletal elements of the face. (Recall the Neural Crest Cell Migration Model that we discussed in the context of Figure 5.1.) Through some very fancy biotechnology experiments, a team of biologists has shown that the formation of fully functioning neural crest cells appears to be tightly tied to how well Gene 2 functions. If Gene 2's instructions are being "read" at full capacity, there are many fully functioning neural crest cells migrating. If the instructions in this gene are scarcely ever "read," fewer neural crest cells will migrate into a given skeletal part of the face.

Like Gene 1, Gene 2 is a master gene: the protein encoded by Gene 2 directs a whole other set of genes that affect the shape and thickness of the bones of the face. It is like an orchestra leader, directing other genes when and where to be active. This master controller directs the migration of neural crest cells so that each plays the appropriate role in shaping the face. So that is how the face is shaped.

The significance of the face in communicating with other people can hardly be overstated. We can learn much about what a person is thinking and how they are feeling simply by watching their facial expression. It is the finely tuned muscles attached to delicately shaped facial skeletal elements that facilitate the versatility we need to use our face to communicate a wide variety of emotions.

Here is an example. The reduction in the brow ridge is one of the distinctive differences between our species and Neanderthals. In a recent paper, three British researchers have shown that this reduction would have enabled significantly greater flexibility of using our eyebrows to reflect one's mental state:

> Mobile eyebrows without the constraints of a pronounced brow ridge allow subtle affiliative emotions to be expressed, such as the rapid "eyebrow flash,'" lasting around one-sixth of a second, found cross-culturally as a sign of contact readiness and recognition. In contrast, a slow eyebrow raise is a sign of surprise and, in particular, social indignation. The facial expression of sympathy—shown by pulling eyebrows up at the middle—has the advantage of removing need for the direct contact used to express sympathy in chimpanzees. Subtle dynamic movements of eyebrows are also a key component of identifying trustworthiness as well as subtle indications of deception. Any constraints on muscle movements in the supraorbital region affect emotional expressions and, in turn, social relationships; for example, individuals who receive a cosmetic procedure (Botox) that reduces muscle activity in the forehead (and so affects

> eyebrow movement) are less able to empathize with and identify other's emotions.[28]

Skeletal alteration of the face is only *one* of the many aspects of the suite of characteristics that has revolutionized the existence of our species and made us into masterful communicators with each other, but it is certainly an important one. The importance of communicating emotions to each other nonverbally may be best illustrated today by considering our interactions with infants. Long before we communicate with words, facial expression is the way we express our thoughts and feelings. Indeed, beginning in their first few months, we engage infants in two-way proto-conversation with a chatter of sounds and loving gesticulations of the face. This universal "language" triggers the mutual release of the hormone oxytocin,[29] the important signaling molecule that helps to cement the parent-child love bond. Before there are words expressing parents' thoughts to the infant, there is this language of facial expression.

Extending this back in time, it would not be surprising if there was a parallel to this—a staging ground—where facial communication served a foundational role in the origin of spoken language as our species came into being. Likely, as members of our species became increasingly communicative with each other, facial expressions would have served as a sort of proto-language, binding our ancestors to each other as they communicated in various other ways, including gestures, music, and eventually, a growing vocabulary. Thus, any genetic changes that alter the facial skeleton in a manner which enhanced the versatility and subtlety of facial expressivity would facilitate communication. And any genetic change that would increase the efficiency of person-to-person communication would enhance the common mindset and cooperation that most investigators believe was the secret to our success.

So, we have defined Gene 2's central role in facial expression—it is a master regulator of the skeletal components of the face. But the question remains, have changes in Gene 2 been taking place as our species came into being and how would we know? To answer that question, we first must establish how we know that Gene 2 affects the migration of neural crest cells.

Mammalian embryos develop in the dark depths of the mother's uterus far out of the range of our best microscopes. Neural crest cells, each with a diameter corresponding to about 1/100 of the thickness of a dime, sit just above the part of the embryo that becomes either the spinal cord or,

28. Godinho et al., "Supraorbital Morphology."

29. Tomasello et al., "Understanding and Sharing." See also Hare and Woods, *Survival of the Friendliest*, 73.

at the embryo's head end, the brain. How could we study the formation and subsequent migration of such cells deep in the mother's body? Well, it is possible to work with neural crest cells outside of the body in a small dish, and under certain conditions investigators can get them to migrate much like they do when they are in the embryo. It is also possible to turn the frequency of reading Gene 2's instructions up or down, much like one can change the speed of reading an audiobook. When the frequency of reading of Gene 2 was slowed down, the rate at which neural crest cells moved over the surface of the dish was altered accordingly. So, Gene 2 really does impact neural crest cell migration.[30]

Showing that Gene 2 affects migration of neural crest cells is consistent with the "softening" of the facial features of Williams syndrome persons who have only one copy of Gene 2 instead of the normal two. But that just leads to another question. Most neural crest cells do not mature into parts of the facial skeleton; they are made into a host of other structures as well—from the adrenal gland, to teeth, to pigment-producing cells, to cells of the peripheral nervous system. How do we know that the neural crest cells affected by Gene 2 become part of the facial skeleton?

We have already learned that Gene 2 is a master gene; it makes something called a transcription factor, which can control the expression of other genes, sort of like a master temperature control regulator can differentially regulate the sub-thermostats in different rooms which are under its direction (see Figure 5.2 again).

The transcription factor produced by Gene 2 has been purified and it is possible to track which "sub-genes" it binds to and controls. From this, we can tell what *sorts* of genes it controls. Can you guess what sorts of "sub-genes" are controlled by Gene 2? It controls a set of genes that medical geneticists have already identified. When mutant, these "sub-genes" result in abnormal skeletal features of the face. Thus, Gene 2 not only affects neural crest cell migration, but more specifically it affects the neural crest cells that give rise to the facial skeleton.[31]

30. One review of the work summarized the results this way: "In this study, researchers from the team led by Giuseppe Testa used *in vitro* (i.e., in dishes) models of Williams syndrome with [neural crest] cells derived from the skin. Results showed that [Gene 2], which lies in the region of the genome causing Williams Syndrome, controls neural crest cell behavior. Lower levels of [Gene 2] resulted in reduced neural-crest migration, and higher levels produced greater neural-crest migration." Phys.org, "First experimental genetic evidence."

31. It likely affects the migration of other neural crest cells as well, but the main point here is that among those which it affects are those that become part of the facial skeleton.

As a master gene, Gene 2 is not restricted to affecting the expression of sub-genes that affect the facial skeleton. Researchers have found that many of the sub-genes controlled by Gene 2 are correlated with domestication in dogs, cats, horses, and cattle. In other words, these are genes which are expressed differently in the domesticated forms of these animals as compared to how they are expressed in their non-domesticated counterparts in the wild. Not only that, but Gene 2 itself is expressed differently in domesticated species than in it is wild counterparts. But has this domestication-related master gene changed in our lineage compared to a cousin species like Neanderthals? Is it possible to test for that?

Remarkably, since DNA is sometimes preserved in ancient bones, it has been possible to isolate DNA from the fossil material of Neanderthals and to decipher its coding instructions—the coding instructions for building the body of a hominin that went extinct forty thousand years ago.[32] Keep in mind that Gene 2 is a master gene which functions to adjust the "thermostats" of sub-genes we know influenced the process of domestication in various animals. So, the question is, have those thermostats been altered more in our lineage, compared to that in Neanderthals? The answer is an unequivocal yes.[33] Indeed it is possible to determine the actual "thermostat setting" for Gene 2 in Neanderthals compared to our own.[34] In essence the "thermostat" is turned down in our species compared to the original "setting" and it is likely that this played a key role in bringing about the "softening" of our facial skeleton compared to that of Neanderthals.

In other words, it now appears that changes to both the master regulator gene and its set of sub-genes involved in facial construction have been altered in our lineage, and that this set of changes parallels a similar set that took place in animals when they were selected for increased docility on their own road to domestication, thousands of years ago.

What is especially remarkable about all of this is that these alterations in Gene 2 increased (on average) the number of offspring. Geneticists call this phenomenon "positive selection." It is beyond the scope of this book to explain the technical details, but in essence what this means is that it is

32. The person most responsible for developing this astonishing technology, Svante Pabo, was recently awarded the Nobel Prize.

33. Zanella et al., "Dosage Analysis of the 7q11.23."

34. Some readers will know that the methylation state of a gene gives us information about the degree to which the activity of a gene is repressed. The methylation status can be determined on ancient DNA and this is what I am referring to here as the "thermostat setting."

possible to show when there are certain regions of a chromosome that carry versions of genes that have clearly been beneficial to reproductive success. The versions humans now have of Gene 2 and many of the sub-genes controlled by it are in chromosome regions that show telltale signs of having benefitted the reproductive success of our species.[35]

Can we say that our species has been domesticated in a manner that parallels the genetic, anatomic, and physiological changes in the animals that became increasingly docile and friendly? Michael Price, a reporter for the journal *Science*, interviewed the Harvard anthropologist Richard Wrangham to get his take on the experiments just discussed above:

> "The study is incredibly impressive," says Richard Wrangham, a biological anthropologist at Harvard University who was not involved in the new work. It's "a really beautiful test," he adds, of the long-standing idea that humans look so different from our primate ancestors precisely because we have become domesticated.[36]

Not all would put it as strongly as Wrangham, but most would agree that something akin to a domestication-like process has influenced our genetic makeup. Although it is doubtful, I think, that there is a perfect parallel between what happens to other mammals when they become domesticated and that which has happened to humans, what *is* clear is that the Sapiens genetic makeup has undergone a dramatic set of changes that have led to vastly improved communication skills and increased cooperation. So noteworthy is this effect that Hare and Woods chose to entitle their important book *Survival of the Friendliest*, and Wrangham, focusing on our species' amazing ability to get along so well with each other, titled his book *The Goodness Paradox*.[37] Hare and Wood point out that although there is no parallel to the fox experiments that allows one to document step-by-step what happens as a species is selected for nonaggressiveness, nonetheless "domestication fossilizes."[38] By this they are referring to the gradual changes in the Sapiens skull, which are thoroughly consistent with a domestication-like process having occurred.

35. Zanella et al., "Dosage Analysis of the 7q11.23."

36. Price, "Early humans domesticated themselves."

37. Both sets of authors acknowledge there is a dark side to humanity as well, a subject we will discuss in chapter 7.

38. Hare and Woods, *Survival of the Friendliest*, 67.

INCREASED "FRIENDLINESS" AND THE SUCCESS OF OUR SPECIES

As we have already described on page 139, more than eighty thousand years ago, a process leading to the increased gracility of the human face began. Domestication (or at least something like it) was underway. Hare and Wood go on to describe the ramifications:

> Domesticating a wolf brain or an ape brain is impressive. But when you domesticate a human brain—this is when the real magic begins. An ultracultural species is born. A unique type of friendliness must have evolved in our species that allowed for larger group sizes, higher population densities, and more amicable relations between neighboring groups that in turn created larger social networks. This encouraged the transmission of more innovations between more innovators. Cultural ratcheting went from slow and sporadic to fast and furious. The result was exponential growth in technology and the emergence of behavioral modernity.[39]

They summarize what they consider to be the self-domestication of humans by writing:

> Self-domestication gave us a superpower, and in the blink of an evolutionary eye, we took over the world. And one by one, every other human species went extinct.[40]

Friendliness, together with the human supercharged brain, say Hare and Wood, unleashes the power of rapid cultural change to a degree never seen in the animal world.

Recall Varki and Brower's proposal discussed extensively in the last chapter. They stressed that it was "reality denial" with its accompanying denial of the reality of death which led to the full theory of mind (ToM). This in turn, they suggested, was the basis of the success of our species. The proposal central to this book is that it is in fact the activity of God's Spirit and the ongoing receptivity of community members in one or more populations that facilitated crossing of the Rubicon to full ToM.

Hare and Woods ask why bonobos never advanced to a full theory of mind. Recall that bonobos are much less aggressive than their chimpanzee cousins. Given their penchant for using sex and grooming to relieve tension, they are sometimes referred to as the hippies of the ape world. Indeed,

39. Hare and Woods, *Survival of the Friendliest*, 66.

40. Hare and Wood, *Survival of the Friendliest*, 65–66.

several investigators have proposed that bonobos are the epitome of self-domestication. Given these domestication-like changes and their genetic proximity to us, Hare and Wood ask, "Why aren't bonobos driving cars?"[41]

They, like Varki and Brower, go on to address the question of how humans made it across the Rubicon to a full ToM, but their explanation is different. They suggest that Sapiens were able to inhibit the neural circuitry associated with emotions—to plot out a best course of action that is not based purely on a "gut response." They outline some of the neurological changes in the brain that allowed our species to be better at putting the brakes on an emotional response to social situations that evoke anger, thereby creating a greater level of tolerance and ultimately friendliness. This is what is of course needed for expanding social networks and this in turn provides clear advantages to genetic changes that would lead to an enhanced theory of mind, all of which is associated with what they (and others) call the self-domestication process.

> It is this self-control combined with reduced reactivity that creates the human-specific adaptation for unique forms of social cognition.[42]

Like the "reality denial" hypothesis of Varki and Brower, the "self-domestication/survival of the friendliest" hypothesis of Hare and Woods assumes that the process of human evolution is a natural process; it assumes there has been no transcendent Spirit gently nudging individuals towards actions that are wise and loving. A Christian theological approach to the question says that neither explanation is complete. Our species developed a full theory of mind because individuals became, through the action of the Spirit of God, aware there was a deeper reality to their existence. But their journey to that enhanced mentalizing capability also included the growing intensity of interaction with the Spirit of God, increasing their ability to suppress those "gut reactions" of animosity and to begin the process of living into a life of growing wisdom, humility, and grace. All these traits when exhibited consistently in one or more communities would not only lead to domestication-like changes, but they would also facilitate and expand meaningful social interactions. This in turn would lead to the success of Sapiens such that they not only replaced all other hominins on earth, but they became the only species to drive cars, design spaceships, create magnificent music, and produce art that enables our spirits to soar. The "ultracultural species" had been born.

41. Hare and Wood, *Survival of the Friendliest*, 58.

42. Hare and Woods, *Survival of the Friendliest*, 58.

Culture can be "defined as behavior transmitted by social facilitation and learning from others."[43] By no means are we the only living species that has culture. Other big-brained species like cetaceans (dolphins and whales) and apes have social traditions that are learned and passed on. "Whales," Carl Safina tells us in his beautifully written book about animal culture, "share our air, but play, socialize and live entirely in the sea. All that makes whales *what* they are. Culture, in the form of differing group behaviors, and vocal identities, make whales *who* they are."[44] They live in a world that is culturally rich—but it is whale culture. He documents this by following them in the Caribbean along with a scientist who has spent his entire career seeking to get to know individual whales and the pods in which they live. Living among them is key to understanding their culture.

In another marvelous book, Jennifer Ackerman's *The Bird Way,* we see that even small-brained species like some birds have the equivalent of culture. It can be discovered, she shows us, as one seeks to "live into" their way of life. For example, we know that there are about two dozen species of bowerbirds, and all except one go through a fascinating mating ritual. Male bowerbirds laboriously construct structures out of sticks. They decorate them with colored objects that serve as the backdrop, and then having completed their massive building project, they seek to attract a female by performing an extremely elaborate courtship dance in front of the bower. Most of this is done by instinct, genetically programmed, and thereby not cultural. However, in some cases, young male birds are known to watch the more senior members of the community, learning from them how best to construct and decorate their bowers. As they grow older, they become experts at constructing bowers that bear certain trademark characteristics of their own geographical area. Much as cities and regions within a given country have their own language dialect, birds also have their own geographical trademarks (including song dialects), revealing a culture specific to the world in which they live.[45]

Culture and cultural evolution emerge from the biology of the species, and it is passed on from generation to generation, in parallel with genes as they are passed from parents to offspring. Cultures change, just as genes change. Culturally induced characteristics change much more rapidly, but there is a sense in which those changes are inherited much like genetic alterations are. Thus, culture transmission can over time play a central role in "shaping the behavior, ecology and even the biology of populations." But

43. Fuentes, "Setting the Stage," 14.

44. Safina, *Beyond Words,* 33.

45. Ackerman, *Bird Way,* 236.

humans take this and supersize it to a fantastically new level.[46] We have discussed this earlier in the context of feedback loops established with the manufacture and use of stone tools, a cultural phenomenon. The habit of making these tools (a culturally transmitted phenomenon) itself affected which gene variants would benefit the population. Gene variants that improve dexterity would be beneficial, as would gene variants that improved effective communication, thereby teaching the next generation how to make good stone tools. There is a sense, then, in which the cultural practices shape the evolution of the species. The theological richness of that statement cannot be overstated, as we shall soon see.

It is easy to envisage that cultural practices in a particular community or set of communities include those which lead to an increased sensitivity to the Spirit of God. Since we cannot go back to experience firsthand what those communities were like, it is not possible to get specific about the practices that may have led to increased sensitivity to the Spirit's promptings. There is something "contagious," however, about goodness that is the antithesis of another contagion—that of evil. The beauty of goodness can be passed on culturally generation to generation, in a way that shapes a community. First John tells us that "if we love one another, God lives in us, and his love is perfected in us."[47] Of course we are aware that this was written to a specific group of early Christians. But it still applies to us in today's culture despite the vast difference of our world compared to that of two thousand years ago. Similarly, surely those words would have applied to those super-ancient cultures scores of millennia ago in Africa. As some of those small groups chose to center their cultures in loving relationships, God would have "lived in them," too, just like John told the community to which he was writing so long ago. And as that happened—generation after generation—it would have led to domestication-like changes in our species. The acquisition of domestication-like genetic changes is exactly what one would expect if the God of Christian theology is real and if our primate genetic makeup were being transformed because a population (or set of populations) was living into the love of God for a period of time.

Let's examine one more aspect of the animal domestication process that can be observed through scientific investigation. Once the selection process for nonaggressive foxes was initiated, there were noticeable genetically based changes within a dozen or so generations. Researchers have been amazed by the rapidity of the process. Mutations are exceedingly rare events, so there is no way the foxes were becoming tame because of mutational changes in

46. Fuentes, "Setting the Stage," 14.

47. 1 John 4:12 (NRSV).

genes. It did not require a period of waiting for the right mutations to come along to make them less aggressive. Rather, their personalities underwent a change because previously existing genetic variants were combined in a manner that *together* led to the nonaggressive characteristics. There was a rapid "weeding out" of variants that led to aggressiveness. The best model we have so far suggests that the new combination of genetic variants slowed down the formation and migration of neural crest cells.[48] However, for the questions we are addressing here, the exact mechanism is largely beside the point. What is important is how readily the species changed when the foxes were selected for a gentler temperament. The new "fox culture" came about because a new component (human selection for nonaggressiveness) was affecting what they were becoming. Their niche had been reshaped by outside influence. Still today, they have not yet become a new species—that would take much, much longer—but if the nonaggressive niche were to continue for centuries and there was almost no outbreeding, they would eventually become a new friendly species of fox.

When mammals in general have been successfully selected for a tame disposition, the result is a fairly specific suite of genetic changes that produces a characteristic set of physical and physiological attributes. What that means is that these changes can be considered a sort of trademark of what happens to the mammalian body in response to consistent multigenerational selective pressure that lowers aggressiveness. In the case of humans, it is important not to get sidetracked by whether or not we should call it domestication. Certainly, given the uniqueness of the human mind and the unprecedented manner in which culture affected our development as a species, we ought to expect some differences in the "domestication-like" process if it exists in humans. The point is, though, that a suite of the same genetic changes that occur in the process of domestication of other animals *is* associated with changes that have occurred in our lineage as well.

48. It would be easy to get technical here, but that would defeat the purpose of the book. Think of it this way: there are thousands of genes which affect height in humans. When two persons of average height have a tall child it is because the new and unique combination of genes they inherited from their mother and father led to that taller stature. Likely the increased height (compared to each parent) was not due to new mutations (see chapter 2). Rather the increased height was brought about by the specific assortment of preexisting "tallness" genes that they received from each parent. That together with the likelihood that they *didn't* receive certain "shortness" genes would have resulted in their becoming taller than either parent. (Remember, we only get half of each parent's genes.) It would work the same way in the fox experiment described above except the trait being considered is nonaggressiveness (and not height). After only a small number of generations of breeding for nonaggressiveness, the fox population began to change because genes causing aggressiveness were being eliminated and genes for nonaggressiveness were being selectively retained.

Those changes have not occurred in the lineage leading to Neanderthals. Joseph Henrich, chair of Harvard's department of human evolutionary biology, writes, "Natural selection shaped our psychology to make us docile, ashamed at norm violations, and adept at acquiring and internalizing social norms. This is the process of self-domestication."[49] Geneticists have shown that not only have there been changes in some of the same genes associated with tameness in animals, but there has actually been a "positive selection"[50] mechanism at play to facilitate the increased frequency of certain genetic variants. Some process has been doing this—changing the genetic make-up of our species in a manner that parallels to some degree the domestication process. Are we unconsciously doing this to ourselves, as Henrich implies?

Science, with its methodological naturalism premise, has no choice but to propose that there was no influence outside of the natural world bringing about these "domestication-like" changes. To propose otherwise would be to abandon the rules by which science operates. On the other hand, if we consider the all-important words of Alister McGrath, we are brought into a whole different realm:

> Christian theology offers an approach to nature that is grounded in its empirical reality, but transcends the limits of the empirical. It offers us theoretical spectacles, which allow us to behold things in such a way that we are able to rise above the limits of the observable and move into the richer realm of discerned meaning and value. In doing so, it does not descend into fantasy, but makes warranted assertions that are grounded in its deep and rich Trinitarian vision of God. The natural world thus becomes God's creation, bearing the subtle imprint of its maker.[51]

I am suggesting that there *was* "outside" influence in bringing our species into existence and that influence became stronger as the minds of our forebears became "a particularly intense locus of divine action because of its remarkable complexity and relational and volitional capabilities."[52] I am not suggesting that all Sapiens communities responded, only that one or more did, and as they continued to respond to the Spirit's leading, significant genetic change took place.

49. Henrich, *Secret of Our Success*, 319.

50. Explaining how we know that positive selection has taken place for certain genetic variants is beyond the scope of a book like this. Some readers may be aware of a phenomenon known as linkage disequilibrium and they will be aware that this can be a clear sign that a genetic variant in a specific chromosomal segment has been conferring a distinct advantage in the somewhat recent past.

51. McGrath, *Darwinism and the Divine*, 289, italics in original.

52. Ritchie, *Divine Action and the Human Mind*, 346.

Here, with our "theoretical spectacles" in place, are what I propose are the basic principles that brought our species into existence:

- Principle 1: God is love; communities where love was prevalent functioned more harmoniously that those where it wasn't.
- Principle 2: Because God is love, communities where love flourished would have been drawn into the presence of the Spirit of God. (Even when their conception of God was rudimentary or even non-existent.)
- Principle 3: Communities that lived in harmony with the Spirit of God and each other would have thrived, increasing in population size faster than those that did not.
- Principle 4: At least one community lived like this for a succession of generations. (At least for several hundred years).
- Principle 5: Genetic variants which enhanced cooperation would gradually increase in such a community.
- Principle 6. The Spirit of God, in partnership with humankind, would have brought a whole new way of being into existence.

A Whole New Way of Being?

One of the most influential scholars in exploring the ramifications of a domestication-like process in human evolution is Terrence Deacon at the University of California, Berkeley. Deacon emphasizes that we are the only species that can think symbolically. Words, for example, are symbols. A word can represent an object, but it bears no physical resemblance to that which it represents. The word "ball" is a symbolic representation of a type of object with specific characteristics. When you read the word "ball" a mental image of that object forms in your mind. The world we inhabit is grounded in symbolic representations of all we experience. No animal does this naturally. With much training, we can bring some animals into this world with us. Dogs, for example, can come to understand the meaning of some of our words, but they don't do this naturally. We, alone, live immersed in a world of symbolism. Deacon refers to humans as the "symbolic species" and draws a parallel to how this governs all aspects of human life, just as existence in the air governs all aspects of a bird's life or a water world governs all aspects of the life of a dolphin. Symbols are of comparable significance in defining human existence. The reading of this book is grounded in the use of symbols as are the thoughts you have about it. Although it is appropriate to make a comparison with air (for birds) and water (for dolphins), there is one important difference. For dolphins and birds, the medium that defines their existence is external to their bodies; for humans, on the other hand, it is internal. We frame all aspects of our existence by making mental representations of real things—be it objects, persons, actions, or ideas. They are housed in our mind as representations. We are born with this innate orientation; as babies and toddlers, we quickly become immersed into exploring the meaning of things—what they represent and how they are interconnected. Deacon says that we have "essentially become symbolic savants. We find ourselves intuitively and irresistibly 'looking beyond the surface' for hidden meanings."[53]

So important is this new dimension to our existence that, as Deacon describes it, "Biologically, we are just another ape. Mentally, [however], we are a new phylum of organisms."[54] In a fascinating book chapter, Deacon makes an analogy between what we biologists recognize as one of the most significant events in the more than three billion years of life on earth: the formation of the first eukaryotic cell. This is not the place to get into details of cell structure. Suffice it to say, however, that that single event that happened about 1.5 billion years ago occurred as a result of a partnership formed between two previously existing disparate organisms—one a eubacteria and the other an archaebacteria. It was this single amalgamation event that likely made macroscopic life possible. Deacon thinks that the origin of symbolic thinking in our lineage (and only in our lineage) is *that* revolutionary to the history of life on earth.

For us, there are now two components to our existence, the first of which is grounded in our biological heritage. We can trace the genealogy of the human body back through its primate, mammalian, and vertebrate heritage through millions and even billions of years to the first cells. But the second component—the distinctive symbol-oriented human mind—began its trajectory much more recently. Deacon calls this our symbolic genealogy, and it is passed on not from within a DNA molecule, but from person to person through social interaction. For example, the name we give to that round object that is fun to throw and catch—ball—does not get passed on through genes; the name (that is, the symbol for it) is transmitted socially. The *capacity* to use symbols is embedded in the brain and is inherited genetically, but the specifics of how we *use* that capacity is culturally transmitted. When communicating through symbolic representation originated is not clear, but most scholars would likely place it within the context of our genus, *Homo*. It likely began in a rudimentary form about two million years ago.

53. Deacon, "On Human (Symbolic) Nature," 145.

54. Deacon, *Symbolic Species*, 23.

Language itself is largely symbolic and it is the medium we use to communicate our thoughts and ideas to each other. But our vocalizations do not trace their origin back to the primitive calls and sounds made through our primate heritage. Here's how Deacon addresses the origin of vocalization:

> [A] distinctive characteristic of humans that sets our species apart from essentially all other land mammals is our capacity to produce complex vocalizations that involve the articulation of the mouth and tongue with precisely synchronized control of the lungs and larynx. The innately produced calls of other land mammals are comparable to human laughter and sobbing, in being stereotypically structured and invariantly linked with specific arousal states. This mammalian limitation is a consequence of the fact that the descending neural control for the tongue, jaw, and facial muscles is distinct from that controlling the larynx and lungs. This is because the muscles of the mouth and face need to be under complex voluntary control in order to be able to adapt moment to moment to the variety of foods that need to be processed. In contrast, pharyngeal, laryngeal, and lung muscles need to function the same way time after time with each swallowing or breath, and need to do so infallibly and automatically. As a result, in the vast majority of land mammals these automatic systems are controlled locally in the brainstem, with little or no cortical input, as part of the visceral-motor system.[55]

Deacon goes on to address the mystery of this new function that is at the forefront of all that we have become as the "symbolic species." Although vocalizations like laughter, shrieks of fright, and cries of anguish are localized in well-specified brain regions—mostly in the deeper, ancient subcortical parts that are used for analogous functions in primates—the vocalizations associated with speech are different. The brain regions associated with the various aspects of generating and processing speech can be shown to be scattered in various regions of the outer highly convoluted cortical brain region. No new brain structures have been developed making speech possible. Instead, parts of the brain that are used for other functions in animals seem to have been usurped and now function to facilitate various neurological aspects of the production and comprehension of language.

What would have enabled this distinctively human genealogy? Deacon thinks it was a domestication-like process. To illustrate this, he turns to what would seem to us to be a most unlikely example, a Japanese finch. This

55. Deacon, *Symbolic Species*, 142.

species was domesticated 250 years ago and has been bred extensively for its color, but only its color, not for its song. Its counterpart in the wild still exists after two and a half centuries, and studies have compared it now to its domesticated cousin that has existed in a separate lineage for these many years. In the wild, its song is highly restricted to a particular vocalization pattern that plays a significant role in attracting female mates. The domesticated variety, however, has not needed this song to attract mates—they are present in the same cage. Thus, that mate-attracting song has played no evolutionarily important role for 250 years. The domesticate has been selectively bred only for its coat color and is not in any competition for mates. Interestingly, the song of this domesticated finch species, no longer constrained to attract mates, has taken on a whole new set of qualities—a much broader suite of vocalizations. Furthermore, brain scans show that it uses a much wider array of brain regions to facilitate its rich vocalization repertoire. In a masterfully delivered lecture given at the University of British Columbia and currently available on YouTube, Deacon proposes that something like this happened to our ancestors as they also became domesticated.[56] Regions of the brain that had previously been used for other activities associated with the old "undomesticated" lifestyle were freed up and could now be co-opted for the expanding capacity for the various aspects of vocalization, language, and symbolization functions in general.

Deacon (and others) stress that as we moved towards becoming the "symbolic species," we gradually changed our way of examining all that we experienced in the world. Symbols, by definition, referred to something else and there was a significant social advantage to being adept at using them and for understanding their meaning. It was not enough, though, to move towards using a collection of sounds to represent physical objects or activities. It was also necessary to have a built-in curiosity about how to connect the sounds to make broader meaning of what the sounds represent. This predisposition to search for meaning—as if there existed a broader invisible sphere in which meaning did exist—led us into becoming beings in search of meaning. We came to expect purposes more real than the objects themselves.[57]

As a result, the transition to a mindset grounded in symbols "left us forever seeking a meaning hidden within a disembodied realm." It even left our ancestors thinking that the mundane physical world around them derived its value from an outside source.[58] We are meaning-making beings; it

56. https://www.youtube.com/watch?v=OT-zZoPMqgI.

57. Deacon and Cashman, "Role of Symbolic Capacity."

58. Deacon, *Symbolic Species.*

comes about through the emergent cognitive revolution associated with a body geared not to flying through the air, or swimming through the water, but one that is specialized for seeking meaning from within a dramatically new form of existence—one embedded in symbolic representations. This—an existence uniquely characterized by the search for meaning—is exactly the property that I think theologian Sarah Lane Ritchie was speaking of when she spoke of the human mind having the qualities one would expect to be an "intense locus of divine activity."[59]

Noted paleoanthropologist and former curator with the American Museum of Natural History Ian Tattersall puts more of an emphasis on what he thinks to be the suddenness of the emergence of symbolic thought, but he ends up at the same place as Deacon.

> The extraordinary properties of the human brain are likely emergent, resulting from a relatively tiny—and altogether accidental—addition or modification to a complex structure that was already, and exaptively, almost prepared for symbolic thought. A small tweak to an existing—and independently viable—structure gave rise to a new form of interaction among the brain's constituents that gave it an entirely unprecedented level of complexity in function.[60]

Fuentes agrees with Deacon and Tattersall about the significance of our emergence as "the symbolic species." This predisposition to searching for meaning within our symbolic sphere shaped the development of human uniqueness. "It is our niche," he writes, and he goes on to say that the development of this niche, complete with its ongoing feedback processes, is central to the explanation of how humans came to be the way they are. It is "not tied to a specific 'modern' human anatomy as previously assumed."[61] It is the specific nature of our mind, unprecedented, and oriented socially in an entirely new manner that is responsible for our development as a species.

As Teilhard de Chardin wrote in *The Phenomenon of Man,*

> In reality, another world is born. Abstractions, logic, reasoned choice and inventions, mathematics, art, calculation of space and time, anxieties and dreams of love—all these activities of inner life are nothing else than the effervescence of the newly formed center as it explodes onto itself.[62]

59. Ritchie, *Divine Action and the Human Mind,* 349.

60. Tattersall, *Masters of the Planet,* 224.

61. Fuentes, *Creative Spark,* 22.

62. de Chardin, *Phenomenon of Man,* 165.

Then he adds,

> Outwardly, almost nothing in the organs had changed. But in depth, a great revolution had taken place: consciousness was now leaping and boiling in a space of supersensory relationships and representations; and simultaneously consciousness was capable of perceiving itself in the concentrated simplicity of its faculties. And all this happened for the first time.[63]

Fuentes recently pointed to a new way of thinking about how humankind came to be. He stresses that our species' openness to the transcendent has "potentially evolutionarily relevant implications" and he goes on to write, "The human niche . . . is not necessarily bounded by material borders. Thus, multiple modes of inquiry, especially those that engage with some transcendent components as a core premise, may be particularly beneficial when asking questions about the human."[64]

So how did all of this come about? In his beautifully written book *The Go-Between God,* John V. Taylor, reflecting on the characteristics of the Holy Spirit, wrote, "[The Spirit] works primarily by generating awareness and communion."[65] Although Taylor was drawing from Scripture to make extrapolations of the work and character of the Spirit in today's world, we can be confident that if it applies to the present and the biblical past, it applies back into the very beginning of our cognitive awareness. And that—grounded as it is in God's providence—is how our species came to be.

63. de Chardin, *Phenomenon of Man,* 169.

64. Fuentes, *Creative Spark,* 15.

65. Taylor, *Go-Between God,* 200.

6

The Origin of Our Species and Providence

INTRODUCTION

THE STORY OF THE creation of our species is one long story thoroughly grounded in the loving providence of God. It is a story that began more than four billion years ago and was begun with us in mind. In the narrative of Genesis 1, creation is repeatedly called "good" as God finishes each stage. Then, once humans are added to the narrative, it is exclaimed to be not just "good," but, signifying its purpose-filled completion, God declares it to be "very good"! And with that, God rests. Biblical scholars tell us that when Genesis 1 was first told to the Israelites, they would have heard it as a temple narrative of the sort that other nations told as they described their gods' temple construction, the place from which those gods would rest and reign.[1] The Israelite hearers of the biblical narrative would have understood that this was *their* temple story: the ordering of the temple of Yahweh, the one true God. The whole cosmos—not a building made by humans—is the dwelling place of God. It is all sacred space, put in place and ordered by God, complete with a declaration of its goodness. But the temples of the neighboring nations always contained physical images of their gods. What about the image of God in the real sacred space—the cosmos? This is where amazing grace comes into the story. God intended that humankind would be God's physical representatives; that was to be our calling and it was for this purpose—deeply embedded in the love of God—that our species was created. As J. Richard Middleton notes, "Creation in Genesis 1 is not

1. See, for example, Middleton, *Liberating Image*, 80, and Provan, *Seriously Dangerous Religion*, 40.

complete (or 'very good') until God created humanity on the sixth day as *imago Dei,* in order to represent and mediate the divine presence on earth."[2] The *whole* earth, with all parts existing in harmonious relationship with one another, is the dwelling place of God. It is not only about us—as much as we are loved and as important as our role in creation is—for God resides throughout. It is all sacred space.

HARMONY

Our species has been on earth for about ten thousand generations—a very long time. By comparison, it has been eighty generations since Jesus walked alongside the Sea of Galilee; this is less than 1 percent of the time since our species came into existence. God's love for humankind did not, of course, begin with the coming of Jesus, nor did it begin with God's promise to Abraham. If there is any one fact central to the biblical message, it is that we have always mattered to God. Surely this would also have been true as our species was being brought into existence; we were born out of and into the love of our Creator. Our view of time contrasts sharply with God's. Intriguingly, even that long stretch of 300,000 or so years during which persons of our species went about their daily routine is only a speck (0.01 percent) compared to the length of time since the creation of life itself occurred. Throughout this vast period, the Spirit, the presence of God's love in creation, has always been active. All that has come to be has been brought into existence through the Spirit's ongoing activity. As John V. Taylor expresses it, "If we are to think of a Creator at all, we are to find him always on the inside of creation. And if God is really on the inside, we must find him in the processes, not in the gaps."[3] Jürgen Moltmann tells us in *The Spirit of Life* that God's Spirit is "the power of creation and the wellspring of life,"[4] and in *God in Creation,* he writes that the whole cosmos is "effected through God the Spirit, and exists in God the Spirit, it also moves and evolves in the energies and powers of the divine Spirit."[5] Examining early church history, Rowan Williams writes that the early church fathers at Chalcedon (451 AD) declared that "God's action works not by displacing but by intensifying from within."[6] The fact

2. Middleton, *Liberating Image,* 87.

3. Taylor, *Go-Between God,* 28.

4. Moltmann, *Spirit of Life,* 35.

5. Moltmann, *God in Creation,* 212.

6. This statement is from a summary of the Council of Chalcedon by Williams, *Christ,* 69. Here is a little more of Williams's summary:"God's action works not by displacing but by intensifying from within the capacity of created agency. In a universe

is that as a biologist, I am astounded by the level of biological insight (see the footnote) with which those early church fathers addressed the nature of divine interaction with creation as a whole. And all of this is consistent with the wonderful poem of God's role in nature, Psalm 104, and similarly in Job with the declaration that if God "took back to himself his spirit and his breath, all flesh would perish together."[7] One of the first things I learned in biology 101 is that ecosystems are characterized by interaction. Organisms do not exist as isolated units. They are part of an interwoven fabric—an ecosystem—with the living and nonliving components always in a state of flux. Openness to interaction is built into the very nature of creation.

Williams describes the analogy between the interaction which exists within the Trinity between Father and Son and the loving relation between Creator and all of creation. The statement requires a careful reading, but it is a beautiful concept and one well worth pondering meditatively:

> Creation's relation to God . . . is grounded in the Son's relation to the Father. Since the Son's relationship to the Father is not a relation of one thing to another thing because God is not finite, but an unimaginably intimate existence in the other, a non-duality that is not a simple identity, we are steered toward a similar model in the relation between Creator and creation. They are not two items that could conceivably be partnered in any list, added to each other, but a relational complex in which one cannot be spoken of without the other. The difference which makes this an analogy, not a pair of parallel instances, is that this relational complex in the case of Father and Son is eternal, entirely reciprocal, and not conceivable as the result of an act of self-determination, whereas in the case of creation it has a beginning, it must be thought of as a free act of divine determination, and is asymmetrical: God would be God without the world.[8]

This love that flows from within the very nature of the Trinity into creation is mediated through the Spirit. What we call physical laws are effected through the Spirit and exist in God. Again, I want to emphasize how Moltmann words it: "The cosmos . . . moves and evolves in the energies and

---

that does not consist of impermeable fixed substances but is a system of interacting finite processes, it is possible to imagine such processes developing in various levels of openness to the fundamental infinite initiative upon which they depend without being bound to a model of Creator and creation as related in a way comparable to mutually exclusive atomistic subjects."

7. Job 34:13 (as translated by Moltmann, *Spirit of Life*, 35).

8. Williams, *Christ*, 218.

powers of the Spirit."[9] Indeed, the laws are consistent, reproduceable, and even measurable because of the regularity of the Spirit's sustaining activity.[10] Calvin University philosopher James K. A. Smith affirms this concept with these words: "[The Spirit's] presence is manifested by God's steady, sustaining care of the universe along the lines of what seem like laws."[11] However, he goes on to say, this does not imply that the Spirit always works in that way or is constrained by these laws. Smith stresses that the universe and natural world "must remain open systems"—open to the Spirit's "ontological surprises." Moreover, Smith expresses discomfort with the terms "*super*natural" and "intervention" because such terms imply the Spirit's absence in nature's ongoing *modus operandi* except in special circumstances. The Spirit is always active in the natural world although "we can think of these special, miraculous manifestations of the Spirit's presence in creation as more intense instances of the Spirit in creation, or as 'sped-up' modes of the Spirit's more regular presences."[12]

Amos Yong, the dean of the School of Mission and Theology at Fuller Theological Seminary and, like Smith, a leading figure in the call for greater emphasis on the Spirit in the science/theology dialog, writes similarly, but with a slightly different twist. Like Smith, he is hesitant about using the term *super*natural. He proposes that the laws of nature are not fixed. Instead, they are simply a reflection of how the Spirit primarily works in the present era. Nature and its laws will function differently in the future age, but there are instances in which the Spirit works "from the future," as the Spirit moves creation into that future. As the Spirit leads the universe towards the realm of eternity, the Spirit's activities will at times work from that future to bring it into its ultimate destiny.[13] This is not as farfetched as it may seem at first. As Christians, we believe that the resurrection constitutes just such a situation.

9. Moltmann, *God in Creation*, 212.

10. The Scottish theologian David Fergusson goes further than viewing the Spirit as "mere sustainer." In words that I think would resonate with Smith and Yong described below, he writes, "Living and active, God strives to generate intended outcomes within material and social processes. This mode of divine activity is not reducible to a mere sustaining and accompanying, though it assumes these. At this juncture, the doctrine of the Holy Spirit requires greater emphasis in a fuller account of providence. God's work is to be characterized pneumatically to convey a sense of its ceaseless and universal scope. Confined neither to an initial setting of created conditions nor to an occasional historical interposition, divine action is spread across space and time." Fergusson, *Providence of God*, 237.

11. Smith, "Is the Universe Open for Surprise?," 891.

12. Smith, "Is the Universe Open for Surprise?," 892.

13. Yong, "Nature Laws and Divine Intervention," 962.

Like Rowan Williams, the recently deceased and much-respected theologian John Webster helps us put the activity of the Spirit in creation into the context of the divine essence as a whole:

> The Holy Spirit is Lord and giver of life, creation's perfecting cause. Creation is distinctly assigned to the Spirit in that he is the divine person by whom created things are brought to their proper end. By the Spirit's motion, creatures fulfil their natures, and their particular mode of being—their *esse-ad*—reaches its term. Originated by the Father and given form by the Word, created things are stirred by the Spirit and impelled towards their good, namely the life which is theirs in the particular mode of God's gift.[14]

All these individuals—Taylor, Moltmann, Williams, Fergusson, Smith, Webster, and Yong—are theologians, not biologists.[15] Speaking as a biologist (and not a theologian), I resonate with their way of thinking. Here is an example as to why this discussion of the nature of the Spirit's work in creation is helpful as we examine the history of life on earth. In a marvelous little book, the evolutionary biologists John Maynard Smith and Eors Szathmary discuss eight significant transitions in the history of life, each of which took the living world's complexity to a new level.[16] From molecules to cells to multicellular organisms and beyond, each transition happened through cooperation among previously independent existing entities. This provides clear testimony to just how significant cooperation has been in stimulating the emergence of new life forms on earth. Indeed, as Harvard

14. Webster, "Trinity and Creation," 8. Webster also notes: "Of each divine work we need to say (a) that it is *absolutely* the work of the undivided godhead; (b) that each person of the godhead performs that work in a *distinct* way, following the manner and order of that person's hypostatic existence; and (c) that particular works may be assigned *eminently* to one person, without rescinding absolute attribution to the undivided Trinity and without denying that the other two persons also participate in that work in the distinct modes proper to them. For the definition of the person to whom a work is eminently assigned includes that person's relations to the other persons and to the single divine essence; appropriation is not individuation." (Webster, "Trinity and Creation," 16.)

15. I recognize that these theological points may sound somewhat speculative for those of us used to thinking more concretely; we who are scientifically inclined are used to having specific experimental data to support our views on cause and effect. Systematic theology does not work that way. But it is important to appreciate that this theology is thoroughly grounded in rigorous logic and in the logic of some of the most profound thinkers of the past two thousand years. We are in interdisciplinary territory here. Science and theology resonate even though at times the specifics must be held in a somewhat tentative fashion.

16. Maynard Smith and Szathmary, *Origins of Life*.

evolutionary biologist Martin Nowak has said, the two main driving forces in evolution—mutation and natural selection—are not sufficient to explain the history of life. In addition to those two "we need a third ingredient, cooperation, to create complex entities from cells to societies."[17] Several years before these works brought the concept of cooperation into biology's limelight, Moltmann wrote that Christian doctrine "views creation as a dynamic web of interconnected processes. The Spirit differentiates and binds together. The Spirit . . . leads living things and their communities beyond themselves. The indwelling Creator Spirit is fundamental for the community of creation."[18] Could it be that this third "ingredient" of which Nowak speaks is an aspect of that which we would actually *expect* to occur, given our theological understanding of the Spirit of God?

For each of the eight major transitions that Maynard Smith and Szathmary document, the basis of the success of the transition is that components are drawn into a coordinated whole, made more efficient by functioning together, rather than separately. By no means are these eight transitions the only examples of such cooperation. Rather, this spirit of intertwined relationship is a principle at work throughout living systems. To illustrate the generality of the principle in the operation of a single cell, we can think of the cooperation between each of its parts in the following way:

> The facts that the parts compose the structure of a living cell, does not imply that the properties of life can be reduced to those of the single components. The properties of life are emergent properties which cannot be reduced to the properties of the components . . . Emergence assumes its real value at the level of properties, and its very notion is based upon the proposition that the emergent properties cannot be reduced to the properties of the parts.[19]

The least complex components in a biological community are the single-celled bacteria. They are known as prokaryotes based on their simple, non-compartmentalized structure, and the small size of their cells.[20] Bacteria interact with other living organisms by modifying and exchanging nutrients in their environment, thereby promoting balance and stability. They,

17. Nowak and Highfield, *Supercooperators*, 14.

18. Moltmann, *God in Creation*, 109.

19. Capra and Luisi, *Systems View of Life*, 133.

20. Technically, prokaryotic organisms are divided up into two classes—the eubacteria and the archeobacteria, or Archaea. There are substantial differences between the two types, but that need not concern us here. Prokaryotes are generally very small, but there are a few cases of giant bacteria.

along with the entire community of life, help to keep "selfishness" in check in a manner that builds and maintains a healthy ecosystem. Indeed, we now know that different bacterial species sometimes exchange genetic information with one another, allowing them to obtain beneficial genes from other species, thereby sometimes allowing others to successfully adapt to newly arising environmental challenges.[21]

For more than two billion years, the biosphere consisted only of prokaryotic cells. Then suddenly, as mentioned in the previous chapter, 1.5 billion years ago, one of the most important events in the history of life took place: One type of prokaryotic cell took up residence in another type and they formed a symbiotic relationship that became permanent in their descendants. The smaller of the two partners went on to specialize in energy production, whereas the larger member took care of general metabolic needs. That partnership proved to be enormously successful: the cells of your body are likely direct descendants of this single ancient event. Indeed, all living organisms that are big enough for you to see (and many that you can't as well) are directly descended from this event.[22] All complex organisms are derived from a union between two species that were able to do so much more when they worked together compared to what they could accomplish on their own. So, cooperation between units is not only at the heart of how a cell functions, or how cells interact to comprise the web of interacting cells in a microbial ecosystem, but also at the core of all complex life on earth, and eventually, of course, to us.

Within all multicellular organisms, the parts of which they are comprised function together as an integrated whole and not as independent units. The study of how parts work together for the good of the emerging whole is called systems biology, and it is summarized nicely as follows: "Organs are subject to the whole-body systems; tissues and cells are subject to the organs of which they are parts; molecules are subject to all the higher levels. And all of these are subject to the sociology of interactions of organisms with each other and with different species. That is what life is about."[23]

So, the harmonious relationship between components is a central theme in the maintenance and evolution of life at the cellular and organism level, but it extends well beyond that. In a highly moving book about a world we hardly know exists—the world of mosses—Robin Wall Kimmerer writes of the interaction between rocks and the various species of mosses that gradually wear stones away, thereby releasing the storehouse of minerals

21. See Lane, *Vital Question*, 82.

22. Lane, *Vital Question*, 39.

23. Noble, *Dance to the Tune of Life*, 181.

that will nourish and sustain new life.[24] Each moss species is specialized for a particular micro-environment and it will make its home among the other species, depending upon the amount of light in one spot versus another, the water availability and desiccation dangers associated with the various nooks and crannies, and presence or absence of other organisms. With each moss species playing its environmental role, the community's actions and interactions are a form of poetry, Wall Kimmerer tells us. As single boulders are worn down grain by grain through the eons of time, the various moss species each play their part in a cycle going back hundreds of millions of years. The diverse collection of minerals settle down into the soil and, together, nourish the ongoing dynamic living processes of all organisms in the vicinity. As for the mosses themselves, each species is adapted to its own specific microhabitat as the community of species release those soil nutrients from boulders: "They are known to one another," she writes figuratively, "intimate partners, the mosses know the contours of the rocks. They remember the route of rainwater down a crevice, the way I know the path to my cabin."[25]

On another occasion she contemplates the dynamics of something we just quickly gloss over in our walk through a forest: a fallen log with its wide array of different moss species.

> A close encounter with a mossy log always makes me think of entering a fantasy fabric shop. Its windows overflow with rich textures and colors that invite you closer to inspect the bolts of cloth arrayed before you. You can run your fingertips over a silky drape of *Plagiothecium* and finger the glossy *Brotherella* brocade. There are dark wooly tufts of *Dicranum*, sheets of golden *Brachythecium*, and shining ribbons of Mnium. The yardage of nubbly brown *Callicladium* tweed is shot through with gilt threads of *Campylium*. To pass hurriedly by without looking is like walking by the Mona Lisa chatting on a cell phone, oblivious. Draw closer to this carpet of green light and shadow, and slender branches form a leafy arbor over sturdy trunks, rain drips through the canopy, and scarlet mites roam over the leaves.[26]

She is describing here a community of interdependent, cooperating moss species, which have been living like this in their little communities for hundreds of millions of years. They exist in their own Lilliputian forest, along with scores of tiny insect species, networks of fungal mycelia, and

24. Wall Kimmerer, *Gathering Moss*.
25. Wall Kimmerer, *Gathering Moss*, 5.
26. Wall Kimmerer, *Gathering Moss*, 10.

billions of soil bacteria, all functioning together as a unit to release the nutrients in that fallen log to provide for the next generation of trees.

Speaking of trees, in a similar book Peter Wohlleben takes us up into that forest, giving us a vantage point from which we can visualize coordinated interactions maintaining the canopy sheltering the mosses.[27] Of a well-studied forest in the Harz Mountains of Germany, Wohlleben tells how "nutrient exchange and helping neighbors in time of need is the rule . . . forests are super-organisms with interconnections much like ant columns."[28] Showing how effective they are, he tells how oak trees, having pumped distasteful tannins through their leaf veins in response to an insect attack, proceed to send chemical signals through their root system to neighboring trees, thereby preparing them to defend themselves against their own ensuing attack. Sometimes the communication between trees is electrical, working in a manner much like nerve signals in animals, albeit a much more slowly moving one. The signals travel through interconnected root systems and frequently through a cooperating network of fungal filaments, which join in the partnerships, helping connect neighboring trees in exchange for fresh nutrients from the trees. So, in a sense, the trees are "speaking" to each other, passing on messages about insects, drought, or other dangers. Sometimes the signals are chemical in nature, and they travel through air, not through the soil. For example, acacia trees send out an aerial warning signal to other nearby trees when giraffes begin to nibble at their leaves. In response, the neighboring trees produce a chemical that giraffes find distasteful, causing them to move to a more distant site.

One of my favorite examples illustrating the significance of cooperation comes from some experiments geared towards trying to increase egg-laying productivity in chickens. Two experiments were run in parallel. In Experiment 1, the most productive individual hens were selected for breeding from a set of cages each with about five to nine hens. This was repeated for several generations—always picking out the most productive hen to produce the next generation of hens. So that is Experiment 1.

In the parallel experiment, which we will call Experiment 2, the selection was not for the most productive individual; rather, the investigators had a whole set of cages, but each generation they would only breed from the cage which had the highest egg productivity.

The difference in results between the two experiments was remarkable. In Experiment 1, where they were selecting for the most productive individual, the productivity unexpectedly went down. What emerged in

27. Wohlleben, *Hidden Life of Trees.*

28. Wohlleben, *Hidden Life of Trees*, 3.

those lines were individuals who were good at suppressing the egg laying of other hens through pecking and even murdering their competitors. The technique worked fine for obtaining a line of bullies, but not for producing lots of eggs.

In Experiment 2, where they were selecting for the most productive group of hens in single cages, rather than the most productive individual, the number of eggs laid increased by 160 percent in several generations. By selecting for the most productive group, rather than the most productive individual, there was little pecking at each other. What emerged was a highly peaceful little community of hens and a much higher level of productivity.

What these experiments illustrate is that "group selection" is needed to evolve the traits that enable everyone within the group to thrive. When selecting within a single group, however, selfishness rules. Chickens are domesticated animals. They were domesticated from the red jungle fowl about eight thousand years ago in southeast Asia and have been shown to have a less reactive temperament than their wild, undomesticated counterpart. Still, even though they have been domesticated—and presumably selected for lower levels of aggressiveness—the genes for selfishness are certainly present. I will have more to say about this as it relates to our own heritage a bit later. But for now, note the clear advantage of being a group of cooperating individuals, even though we know full well that a tendency for self-centeredness is lurking in the background and would likely rear its ugly head if the selection strategy was changed.

So how ought we think about the origin of cooperation in the living world as a whole? John Webster cautions theologians that as they consider God-and-creation, it is best not to start with the nature of creation, and then move upwards to draw conclusions about the source of those characteristics as they exist in God. He emphasized that good theology is done the other way around. Thus, with this in mind he wrote, "the material order of Christian teaching about the Trinity in relation to creatures [is] first the divine essence, then the distinction of persons [i.e., within the Trinity], and (only) then the procession of creatures from God."[29] The nature of creation—creation's character—emanates from within God, whose uncreated divine unifying essence is three-persons-in-one: eternal love from the Father, through the Son, and in the Spirit. As a biologist, I am struck by the notion that the harmony that we biologists see in creation (imperfect as it may be) is a reflection (hazy though it may be) of the nature of God as described by theologians in an increasingly refined manner but dating back to early

29. Webster, "Trinity and Creation," 7. (Square brackets added by me.)

Christianity. For example, according to Augustine (as summarized by David Fergusson), creation participates in the divine nature:

> The spiritual, moral, and aesthetic relation of the world to God is maintained by the manner in which the world derivatively possesses the form of the divine likeness. Created beings can participate, albeit in a manner appropriate to their created status, in the form, order, and weight of the divine life.[30]

Augustine, of course, knew nothing of the origin of the eukaryotic cell formed from two cells which became one; he had no way of knowing that from an event where two parts started functioning as one whole, all macroscopic life, including humans, would eventually come into existence. Augustine did not know that the health of a forest frequently depended upon a web of interconnected trees "talking" to and nurturing each other through chemical and electrical signals. Moving on through time, it was almost one millennium later when another of Christian theology's greatest minds, Thomas Aquinas, weighed in. Simon Oliver summarizes Aquinas's views as follows:

> The eternal relations of the Trinity are real relations in the sense that the persons are wholly mutually determinative: the Father is only the Father in relation to the Son and the Spirit; the Son is only the Son in relation to the Father and the Spirit; the Spirit is only the Spirit in relation to the Father and the Son. So the difference between God and creation is not a random and inscrutable difference; it is a participation in, or trace of, the eternal differences and relations of the Godhead. For Aquinas, the emanation of creation from God . . . is an emanation of, or participation in, the eternal emanation of the persons of the Trinity. So God's act of creation is not simply an act of the divine will, impenetrable to reason. It is the expression of the very nature of God himself . . .[31]

Oliver goes on to write that Aquinas believed that "Creation has a participation in God's being. Whereas God's being is simple and one, it is received in creation in many different ways."[32]

The eight major transitions in life's history described by Maynard Smith and Szathmary were each dependent upon separate parts coming together and functioning as a harmonious unit, even though each of the eight

30. This summary of Augustine's views is from Fergusson, *Creation*, 24.

31. Oliver, *Creation*, 52.

32. Oliver, *Creation*, 72.

represents vastly different aspects of the living world. Be it the behavior of molecules, chromosomes, cells, organisms, or groups of organisms, each cooperatively participates in the harmony that characterizes the trinitarian essence of who God is. The principle that Aquinas described from his theological perspective has been borne out by biology. Eight hundred years after Aquinas described it in theological terms, we are only now coming to realize just how realistic his biological insights were.

CONTINGENCY

So, what does theology tell us about *how* God brought creation into being? About this, theology is largely silent. It happened—Genesis 1 informs us—in response to God calling creation into existence, but the Bible (and theology) does not specifically attempt to take it much further than that by providing mechanistic details. Having said that, however, I think the words of John Wesley are helpful here in thinking theologically about the ramifications of that "beckoning call." Wesley once wrote, "There is no place empty of God, and every point of infinite space is full of God . . ." He also wrote that God acts

> by sustaining all things, without which everything would in an instant sink into its primitive nothing, by strongly and sweetly influencing all, and yet without destroying the liberty of his rational creatures.[33]

In thinking about these words of Wesley, the Wesleyan scholar Michael Lodahl, states:

> Since all theological language is necessarily imprecise (whether analogical or metaphorical), we are reminded that there is no simple formula for describing the mystery of God's omnipresence. Like Wesley, we can affirm "the fact" while remaining largely ignorant of "the manner." As he stated . . . "How astonishingly little do we know of God! . . . What conception can we form of his omnipresence? Who is able to comprehend how God is in this and every place? How he fills the immensity of space?"[34]

33. Cited in Lodahl, "Cosmological Basis," 21.

34. Cited in Lodahl, "Cosmological Basis," 21.

God's love is uncoercive. Although there seem to be exceptions, generally God gently nudges and calls forth, but because of God's love, freedom is a component of creation too. Lodahl summarizes Wesley's view on freedom:

> God 'lets there be' that which is not God, but which depends immediately upon God for its continued being—and that not 'from a distance' but indeed 'from within'! How wonderful is love like this![35]

This discussion of God's omnipresence and God's freedom now leads us back into a discussion of the major transitions identified by Maynard Smith and Szathmary and specifically of their second noteworthy observation—that six of these eight major transitions only happened once. Here is how they summarize the ramifications of this:

> There are interesting implications of the occurrence of six unique transitions, together with the origin of life itself, which we also think to have been a unique sequence of events. Any one of them might not have happened, and if not, we would not be here, nor any organism remotely like us.[36]

This is another way of saying that the probability of a species like ours coming into existence through fixed natural laws is near zero.[37] And yet, it happened! Almost all human evolutionary scientists would now emphasize that our arrival as a species has been a highly contingent event. As important as the phenomenon of natural selection has been to the course of life's history, there is no scientific algorithm that can describe the arrival of our species. Ernst Mayr, one of the founders of the modern evolutionary synthesis, wrote shortly before he died, "the role of contingency and chance in natural processes is now almost universally acknowledged."[38] This was also the perspective of Maynard Smith and Szathmary. The nearly unanimous consensus within the field of evolutionary biology is that the arrival of *Homo sapiens*—when viewed through a naturalistic lens—would not have happened were it not for a whole lot of incredibly lucky breaks. Consider how Washington University evolutionary biologist Jonathan Losos words it:

> If any of a countless number of events had occurred differently in the past, *Homo sapiens* wouldn't have evolved. We were far from inevitable and are lucky to be here, fortunate that events

35. Lodahl, "Cosmological Basis," 29.

36. Maynard Smith and Szathmary, *Origins of Life*, 19.

37. See footnote 49 in this chapter for further discussion of this point.

38. Mayr, "Darwin's Influence on Modern Thought," 80.

> happened just as they did. Asteroids, of course, but what other events critically tipped evolution's path in our favor? Who knows how slight a difference in the past—a tree falling on great-great-great-to-the-millionth-degree-grandpa Ernie, a forest fire, a mutation—might have snuffed out our future existence?[39]

In looking at the human lineage, E. O. Wilson writes that that line "veered close to extinction at least once and possibly many times over the past half million years." All that it would have taken, he goes on to say, would have been a severe drought, an alien disease from nearby animal species, and our lineage would have been gone forever in a "geological eye blink." If so: "There would then have followed—nothing. The evolution of the biosphere would have pulled back, never again to produce what we became."[40]

It was the great theoretical biologist Stuart Kaufmann who, prior to Gould, first proposed the thought experiment of replaying the history of life "myriad times" to see what properties of life would emerge repeatedly.[41] Although it will always only be a thought experiment, it is certainly meaningful to consider it carefully as it relates to the emergence of humans. In fact, there are "natural" experiments that really did take place and speak clearly to the likelihood of the arrival of humans.[42]

First, consider the lineages of Old and New World monkeys which have been going their separate ways for about 35 to 38 million years. Today, there are 124 species of new world monkeys, each with distinctive properties such as a flat nose, nostrils that point to the side, and frequently a long prehensile tail that can be a sort of fifth limb used for swinging through the trees. Old World monkeys, on the other hand, have a narrower snout and nostrils that point down. Many have only a short tail or no tail at all. The earliest fossils of monkeys in the New World are about thirty-five million years old. The fact that those first fossils bear distinctive structural features also found in African monkey fossils of about the same time implies they had a common ancestor—an early African species which lived at about the same time implies they both are derived from a single African population. (There were no monkeys in the New World prior to this.)

If they are each derived from a single African population, how did the two groups become physically separated? Although it is true that Africa was physically joined to South America in ancient history, the continental split and subsequent drift took place 100 million years ago—more than sixty

39. Losos, *Improbably Destinies*, 334.

40. Wilson, *Social Conquest of the Earth*, 14.

41. Kauffman, "Self-organization," cited in Dawkins, *Ancestor's Tale*, 583.

42. The five "Experiments" described here are from de Queiroz, *Monkey's Voyage*.

million years before the branch point in the monkey lineage. So how did they get to the New World?

It seems likely to most investigators that something like the following would have taken place: a small number (perhaps even a single pregnant female) was trapped on a huge tropical tree as it floated down an African river (possibly in a massive flood) and then, having been transported in an ocean current, the tree with its clinging cargo arrived a couple of weeks later on the shores of South America. Once that happened, the rest is history: the next thirty-five million years (or so) became an interesting experiment in evolutionary biology. What took place in that lineage over that vast landscape with its millions of square miles of widely diverse habitats and roughly ten million generations? The lineage gave rise to more monkey species—lots of them—but it never gave rise to anything remotely resembling the ape or hominin lineage. In Africa, on the other hand, monkeys gave rise to a wide range of Old World monkey species (including seventy-eight alive today), all the great ape species that have ever lived, and all the dozens of hominin species—including us. In South, Central, and North America, the same lineage only gave rise to more monkeys. Apparently, even starting off with the same "seed species" or one closely related to it, and providing it with similar environments to those which would have existed in Africa, was not enough to give rise to anything that resembles the ape lineage. The whole ape lineage, including hominins, never developed in the new world despite thirty-five million years and vast tracts of varied habitat. Humans (or something like them) are not inevitable—at least not over thirty-five million years.

So that is Experiment #1. Consider the following second "experiment." Just as continental drift led to the breakup of Africa and South America 100 million years ago, similarly Australia separated from South America about sixty million years ago. Early mammals were "on board" and over the coming millennia, the progeny of those mammals—a group of nonplacental mammals—went on to become the mammalian lineage in Australia. So, this is another test run of Kauffman's thought experiment. Remember, sixty million years is a long time! What emerged from that sixty-million-year-long "test run" on this island continent? Nothing like humans, apes, monkeys, or even a new placental mammal body plan ever appeared there—only more mammals with pouches as well as a small number of egg-laying species, each beautifully adapted in highly specific ways to all sorts of habitats.[43]

43. The platypus and spiny anteater have traits that are derived from an even more ancient form of mammal. All other Australian mammals are marsupials and have pouches, except bats. Australian bat species of course are an exception, which, given their wings, proves the rule.

Evolutionary Experiment #3 took place on the land mass we now call New Zealand. New Zealand broke away from the large supercontinent even earlier than Australia—about eighty million years ago. It carried no mammals at the time of its breakaway and no mammals ever arrived on floating trees. So, the only large animals that ever lived on New Zealand were birds and reptiles. Over a period of eighty million years and a geographical area of 100,000 square miles with a vast array of different climate and ecological niches, nothing with the mammalian body plan ever arose there and certainly nothing like monkeys, apes, or humans.

Madagascar provides more data to consider (Experiment #4). With a land area over twice that of New Zealand, it separated from Africa about 130 million years ago and from India close to eighty million years ago. Around fifty-four million years ago a small number of lemurs made it across the 250 miles from Africa to land in Madagascar. Lemurs are primates like us, and we would have shared a common ancestor with them about sixty million years ago. In Madagascar, lemur evolution thrived and eventually gave rise to over one hundred different species of lemurs. Despite fifty-four million years and a very substantial tropical land area with diverse ecosystems and a primate ancestral species, this experiment also yielded no monkeys, apes, or hominins.

Even in Asia and Europe with their vast land area and with plenty of primates present throughout the past sixty-five million years, nothing like hominins ever arose (Experiment #5). True, *Homo erectus* migrated to there from Africa two million years ago and several other *Homo* species, like Neanderthals, may have originated there from within the hominin lineage, but the hominin lineage itself began in Africa and nothing like it originated anywhere else, not even in Asia or Europe.

So that is a set of five experiments that indirectly give us a feeling for the likelihood that something like humans might arise over a period of millions of years and a vast array of diverse habitats.

In 1970, Jacques Monod, a pioneer in molecular genetics, wrote, "Pure chance, absolutely free but blind, at the very root of the stupendous edifice of evolution: this central concept of modern biology is no longer one among other possible or even conceivable hypotheses. It is today the sole conceivable hypothesis, the only one that squares with observed and tested fact."[44] Fifty years later, Sean B. Carroll, one of biology's most respected writers today, wrote approvingly of Monod's words and then added a bit of an update:

> The domain of chance has expanded into realms [no one ever] imagined. As we have learned much more about the history and

44. Monod, *Chance and Necessity*, cited in Carroll, *Series of Fortunate Events*, 8.

> workings of the planet, we have been startled to discover how the course of life has been buffeted by a variety of cosmological and geological accidents—without which we would not be here.[45]

He then goes on to describe the adaptive radiation of mammals, which began to flourish sixty-five million years ago with the end of the dinosaur era. Although mammals had been present for about 100 million years prior to the extinction of dinosaurs, the explosive radiation of mammalian diversity did not occur until immediately after the event.[46] It is now well established that the dinosaur extinction was caused by the arrival of a six-mile-wide asteroid travelling about fifty thousand miles per hour as it hit Earth, creating the most powerful earthquake of all time. An event of this magnitude was almost a singularity: for *at least* the preceding 500 million years (and possibly much longer) nothing like this had happened. The crater, of this asteroid impact, the Chicxulub crater, has been found at the bottom of what is now ocean off the coast of the Yucatan Peninsula in Mexico; the iridium it carried with it in high abundance has been identified as a distinct layer in the earth's rocks at the expected time, and a layer of minerals formed only through high impact are distinctively distributed in the same layer. The impact triggered a cataclysm of volcanic eruptions, wildfires, tsunamis, acid rain, and sunlight-blocking dust. It is now clear that the dinosaur lineage, especially herbivorous dinosaurs, was already experiencing an ecological crisis that created great instability in the food web immediately before the asteroid's arrival.[47] Both events—together exceedingly improbable—provided the opening needed for the mammalian radiation that soon took over the planet. However, even at that, a recent study has shown that the particular site of the catastrophic crash was fortuitous. Geologists estimate that only between 1 and 13 percent of the earth's surface has rocks rich enough in hydrocarbons and sulfur to send a comparable amount of destructive material into the atmosphere. Given the speed of the earth's rotation (one thousand miles per hour), they estimate that if the asteroid had arrived thirty minutes earlier it would have crashed into the Atlantic Ocean and done much less damage. If it had arrived thirty minutes later, it would have crashed in the Pacific, also causing much less harm. In other words, thirty minutes either way and the dinosaurs would still be here, and we would not.[48]

45. Carroll, *Series of Fortunate Events*, 10.

46. Dos Reis et al., "Phylogenomic datasets."

47. Brusatte, "What Killed the Dinosaurs?," 6.

48. Carroll, *Series of Fortunate Events*, 32. Much more could be said about the

Carroll goes on to describe what may well be the other most significant event in shaping the living things on earth in the past sixty-five million years. The development of new techniques has enabled the tracking of climate change and carbon dioxide levels in meticulous detail over this time span. For the first fifteen million years after the asteroid struck, the mean temperature on earth was about 82°F on average, compared to 57°F today. Tropical or subtropical forest covered much of the earth. Indeed, Antarctica itself was a lush subtropical forest. All of this was due to a very high carbon dioxide level (1,400 parts per million) compared to today's 415 parts per million. Then, suddenly (in geochronological time), the carbon dioxide level fell to half, and with the global cooling that resulted, the mean surface temperature dropped more than 10°F by thirty-four million years ago. The temperature continued to plummet until two million years ago. There were times (during ice ages) when the mean surface temperature was only 50°F. Although there are many factors that were influencing this shift, the main one responsible for the overall decline is understood. I mention it because if this had not happened the earth would likely have stayed very hot, bearing little resemblance to the sorts of habitats that gave rise to our lineage.

What was that major factor? The earth's surface consists of a set of about a dozen tectonic plates that slowly ride like rafts over the semiliquid layer of magma and molten rock below. They each move very slowly—only a few centimeters per year, with one exception. Forty miles thinner than any of the other plates, the Indian plate is an anomaly. As a result, it moves more than five times faster than any of the others—about eighteen to twenty cm per year. This might seem insignificant, but its consequences for earth's biosphere are gigantic. Sixty-five million years ago, the Indian plate was about 1,500 miles southwest of where it is today, just off the eastern coast of southern Africa and a little south of Madagascar. It travelled northeast at its "rapid" speed, "crashing" into the Asian continent about forty million years ago. The "crash" created the upthrust that we know today as the Himalayan Mountains and the Tibetan Plateau. The result of this upthrust was the exposure of a vast amount of previously submerged rock, which was now subject to the forces of erosion. This is not a geology book, so I won't go through the details, but this set in motion a series of geochemical events that "sucked up" vast amounts of carbon dioxide from the atmosphere. It was this event

reasons why dinosaurs went extinct, but not mammals. Carroll provides an excellent discussion of this on pages 26–30. Especially pertinent is his discussion of the susceptibility of large land animals to the toxic effects and climate change caused by the asteroid impact. Burrowing animals (like mammals) and those that spent a significant part of their lives in water (like frogs, crocodiles, turtles, and sea birds) were more likely to make it through the disaster.

and all its subsequent consequences that drastically cut the atmospheric carbon dioxide level and resulted in the concomitant severe drop in temperature (global cooling). If the Indian plate had behaved like all the other plates gliding over the earth's surface and moved at a comparable speed, it would not yet have reached Asia; the Himalayans would not yet exist; the carbon dioxide levels would still be sky high; and the earth would still be a hothouse. The ecosystems on earth would have been distributed in a totally different way and many would not even exist. If this collision had not taken place, Carroll writes, "The world's climate would certainly not have changed in the way it did, and the story of life would have been much different." Carroll then goes on to describe the continued changes in climate, including especially the ice age (a once-in-300-million-years event) and how the subsequent changes in climate were of fundamental significance in shaping the direction that our specific lineage took. "We were born, then," he concludes, "out of a rare ice age set in motion by a long-ago geological accident and forged by one of the most unstable and unpredictable climate cycles any mammal has encountered—a series of fortunate geological events."[49]

As his book comes to a close, Carroll takes his worldview to its naturalistic conclusion, assuming, as he does, that there is no other reliable source of information about the origin or our species than science. He quotes from the comedian Ricky Gervais:

> Why are we here? Well, we just happened to be here, we couldn't choose it. We're not special, we're just lucky; and this is a holiday. We didn't exist for 14 and a half billion years. Then we got 80 or 90 years if we're lucky, and then we'll never exist again. So we should make the most of it.[50]

Following that quote, Carroll ends his book with these words: "Tell the truth, treat others kindly, create and for God's sake laugh." As much as I love Sean Carroll as a writer who eloquently describes the exquisite beauty

49. It is important to emphasize that what is being discussed here is the origin of our species, *Homo sapiens,* and the unique form of consciousness we possess. As Simon Conway Morris has demonstrated so profoundly in his books (for example, *Life's Solution)*, the evolutionary process does give rise to similar anatomical structures over and over again through convergent evolution. Similarly, Andreas Wagner (in *Arrival of the Fittest*) and others have shown that this is true at biochemical level as well. However, as Carroll (as well as many others) point out, the arrival of a species having our unique set of characteristics has only happened once and it was by no means an inevitable consequence of the evolutionary process. (Recall Deacon's statement discussed in chapter 5: "Biologically, we are just another ape. Mentally, [however,] we are a new phylum of organisms.")

50. Carroll, *Series of Fortunate Events,* 178.

of living systems, and admire the scientific advances his research group has made in a subdiscipline very dear to my heart—developmental biology of the little fly, *Drosophila*—I think he has dismissed other sources that inform us about reality too quickly. This notion that natural selection—if given many, many chances at a rerun—would almost certainly not produce anything like us may well be among the most theologically significant findings[51] emerging from the sciences today. If so, we can suggest a slight change to the final words of his book: Tell the truth, treat others kindly, create, and laugh out of pure joy because you have been created to participate in the life of the One who created you. The origin of our species is grounded—not in luck—but in the providence of God.

CONTINGENCY IN THE LIGHT OF THEOLOGICAL CONSIDERATIONS

The Bible is one long story of God's providence. As the Old Testament tells the story of the people of Israel becoming established as God's chosen people, or as the book of Acts details the start of the early church, the unfolding events take place within the providence of God. For example, as Genesis draws to a close, Joseph is shown reflecting on how his brothers' hateful act became part of God's providential plan. "You intended to harm me, but God intended it for good to accomplish what is now being done, the saving of many lives," he said.[52] The active involvement of God in Joseph's life is largely hidden, albeit in a manner that is accessible by faith. The other participants—Joseph's father and brothers—would likely not have realized that God was at work bringing about a particular outcome had it not been for Joseph's faith-filled reflection. Still, the biblical narrative makes it clear that the unfolding of Joseph's rise to power in Egypt was wholly contingent on the providence of God.

The narrative of the emergence of the early church in the book of Acts is one long, beautifully articulated story of the providence of God manifested through a series of miraculous events. For example, Paul is visited by the resurrected Jesus on the Damascus Road and that event results in his reversing course from being a persecutor of Christians to being Christianity's

51. Note that I said *theologically* important. Recall the point made earlier in this chapter in discussing the thinking of John Webster: Good theology is done by "starting with the divine essence . . . and (only) then to the procession of creatures from God." I am not equating this with a scientific conclusion about the existence of a creator and will have more to say about that below.

52. Gen 50:20.

most eloquent spokesperson.[53] Similarly, both Cornelius and Peter had almost simultaneous visions from angelic messengers that opened the way for the church to spread out into the entire Mediterranean world.[54]

In an extensive analysis of the biblical view of providence by leading Christian thinkers through the ages, David Fergusson concludes that the biblical portrayal of God's providence is multidimensional. God works, he writes, in various ways to carry out God's purposes:

> On the basis of all this, we might further refine the hypothesis by stating that providence in Scripture is general (it signifies the way the world is made), universal (it is neither occasional nor intermittent but pervasive in all societies through the agency of wisdom), particular (it determines the individual circumstances of people and communities, including Israel and the church, where some patterns and shapes can be discerned), hidden (it requires the discernment of faith and even then is often inscrutable), contested (the presence of evil, suffering and misfortune create obstacles and crises for our more naïve intuitions) and incomplete (an eschatological resolution is promised and anticipated).[55]

Fergusson emphasizes that there are two "bridge concepts" that are important in describing the ongoing creative activity of God: wisdom (Hebrew, *hokmah*) and spirit (*ruah*). "As universal, pervasive and active, divine wisdom permeates the created order,"[56] and "the spirit of God is everywhere present."[57] But he goes on to emphasize the *mysterious* elements of providence. For example, even though the speeches of God at the end of Job point to the grandeur and beauty in creation,

> they also teach that this is not a risk-free world or one that is devoid of suffering and misfortune. Purposes that elude human understanding are suggested . . . The world is created, ordered and constrained but without a tight control or immediate interposition by its Maker to ensure those outcomes that human beings might prefer . . . Divine wisdom requires Job to see a bigger picture, although not one that can be fully comprehended or rendered free of flaw and ambiguity.[58]

53. Acts 9.

54. Acts 10.

55. Fergusson, *Providence of God*, 41.

56. Fergusson, *Providence of God*, 23.

57. Fergusson, *Providence of God*, 25.

58. Fergusson, *Providence of God*, 24.

Similarly, Walton and Longman, in assessing the message in Job, write, "we are [not] capable of devising a sufficiently comprehensive understanding to which we can hold [God] accountable."[59]

As I mentioned above, almost all evolutionary scientists would now emphasize that our arrival as a species would not have occurred were it not for a countless set of low probability, contingent events. Is it not worth the while of these scientists to consider that there might be more to this story than luck? Fergusson writes of the "hiddenness" of providence and that it requires "a discernment grounded in faith." Faith requires a degree of surrender of self-sufficiency; could it be that the desire for personal autonomy is sometimes a reason for the hesitancy of scientists to consider an option other than luck? Perhaps others, like Job, have difficulty seeing beyond the conception of a creator who exhibits "tight control" and "immediate interposition."[60] But the biblical story, when considered in its entirety, does not always portray providence in that way. It seems to me that a sort of tunnel vision is required to assert *with near certainty* that the origin of our species is the result of luck alone.

One of the twentieth century's most influential theologians was Karl Barth. In summarizing Barth's views on providence, Kathryn Tanner writes:

> One should turn to Jesus Christ for one's understanding of a world created and ruled by God rather than draw conclusions about such matters from more general observation of the world and its natural and historical processes. One should not, then, assume that the world exists and search for its cause . . . When beliefs about providence stem *from* Christ rather than move *to* him, providence becomes, not the immanent movement of created powers, but those arrangements of the world that reflect God's prior intentions for the world in Christ, arrangements that reflect God's primary intention to draw the world towards him.[61]

Before the foundation of the world—we are told in the first chapter of Ephesians—God had a vision of what we were to become in Christ.[62] If we want to understand the origin of our species, we need to start with God's vision and not ours. The first chapter of Colossians tells us that Christ was the firstborn of all creation, the visible image of the invisible God. The thesis at the center of this book, *On the (Divine) Origin of Our Species,* is

59. Walton and Longman, *How to Read Job*, 48.

60. See the Fergusson quote in the previous paragraph.

61. Tanner, "Creation and Providence," 111 (emphasis added).

62. Eph 1:4.

that scientific tools are only able to give us a partial picture of reality. Belief in God, as revealed through the life, death, and resurrection of Jesus Christ, completes the picture. It is the love of God that has brought our species into existence—not luck.

As Fergusson writes:

> Christian theology seems heavily invested in the notion of an evolutionary directionality. The appearance of *Homo sapiens* (or something like us) is often read as the inevitable culmination of a long history of increasing complexity in which language, culture, art, science, ethics, and religion must emerge. In an important sense, we must have been intended; hence we are inevitable in the course of evolution . . . According to the cosmic Christology of the New Testament, the human appearance of the Word of God was intended *from the outset* to bring creation to its appointed goal. This suggests an inherent direction in the course of evolution.[63]

If the God of Christian theology exists, there would be, according to Fergusson, a directionality to creation. *Homo sapiens* (or something like us) would be expected. We are here on purpose and not by accident, despite the titles of a couple of recent books on human origins that place the term "accidental species" front and center.[64] Our arrival as a species here on earth was indeed contingent, but its contingency is not based on luck; rather, it is grounded in the very nature of God's character. Rowan Williams writes that creation is most fully itself when it is aligned with the relationship between the Father and Son, a relationship that has existed for all eternity. He goes on to say:

> The logic of creation includes a natural trajectory towards this kind of life-giving responsiveness (i.e., Father to Son). If this is the goal of creation as such, creation must include the possibility of finite creatures who are endowed with love and intelligence after the image of the Son and who are therefore capable of responding to the Creator with something of the conscious joy and generosity which would reflect in finite terms the character of the Son's self-giving to the Father. The focal point, the "gathering point," of the created order is thus this finite love and intelligence, which we recognize as distinctively present in humanity . . .[65]

63. Fergusson, *Providence of God*, 188 (italics mine).

64. Gee, *Accidental Species*; Tattersall and DeSalle, *Accidental* Homo sapiens.

65. Williams, *Christ*, 218. Note: I have added the parenthetical statement (Father to Son) for clarification.

So, Christianity is infused with purpose, grounded in God's love; we were intended—before time began—to live into that love. The coming of humans *was* contingent, but it was a contingency grounded in the eternal plan: "For he chose us in him before the creation of the world to be holy and blameless in his sight."[66]

LIMITS TO THE HUMAN UNDERSTANDING OF PROVIDENCE

Pinpointing the Occurrence of Miraculous Events in Evolutionary History

At this point I want to address two specific concerns related to the human limits of understanding God's providence in bringing about the origin of our species. Here is the first: Can we single out specific events in the history of the evolutionary origin of our species and declare that these single events happened in a miraculous manner, one that supersedes natural laws? I will illustrate this first concern by going back to a key event that likely played a significant role in the success of the hominin lineage—a mutation that almost certainly was historically important in giving rise to the larger brains that came to characterize our genus, *Homo*. We discussed this in chapter 2. Out of about 6.2 billion code letters in DNA, one of them was altered, quite possibly because of a momentary quantum fluctuation in the chemical structure of one DNA base at the instant of replication in a single sperm or egg several million years ago.[67] Mutations occur regularly, so that in itself is not particularly remarkable. However, in this case, that specific alteration triggered a long succession of changes in the metabolism of certain brain cells. In reading one scientific article describing the altered series of events, one has the sensation that they are examining one of those Rube Goldberg cartoons.[68] The result of this step-by-step-by-step set of changes was a dramatic increase in the energy produced within neuron stem cells. What is remarkable is that the entire "Rube Goldberg apparatus" would have had to

66. Eph 1:4.

67. As readers with a background knowledge in organic chemistry will know, point mutations of this sort can occur in various ways. Here is one: The bases in DNA occasionally undergo an instantaneous keto to enol transformation. Since the base pairing can be different when in one conformation rather than the other, if the extremely brief transformation happens during the DNA replication process it can result in the wrong base being inserted.

68. Check out "Rube Goldberg machine" in Wikipedia if you're unfamiliar with this term. Or consider the "butterfly effect"; it constitutes another good example.

have been set up in advance. All that was needed was for the initial highly specific mutation to occur in order to trigger it. As I read the scientific article describing the cascade,[69] it seemed that I was learning firsthand about a miracle, so fortuitous did this complete set of "prepared" events appear. But how would I know? Could it not have been that the cascade already had some other function, and that this mutation was just tapping into a set-up that had another purpose? We just cannot be sure, nor do we know if there might have been some other explanation (such as a couple of almost simultaneous changes occurring by chance). It is not, in my opinion, that miracles have not taken place to bring our species into being. We just do not know enough to pinpoint where and when they occur. The late noted physicist and theologian John Polkinghorne made a point that I think is germane here. He wrote that there are times when one might expect the possibility of miracles:

> It is perfectly coherent to believe that in new circumstances ("the fullness of time") God will do new things. This realization is the basis of accepting the possibility of the miraculous, in particular the central Christian miracle of the resurrection of Jesus Christ, seen as the seminal event from which the great divine act of the new creation has begun to grow. The theological credibility of miracle depends upon being able to exhibit the divine consistency that underlies the novelty, for there is certainly no theological reason to suppose that God must always be boringly restricted to doing nothing new.[70]

The question for Polkinghorne, however, is what constitutes "new circumstances" and what is the "fullness of time"? The resurrection of Jesus is one example he gives. For those who, like Polkinghorne, have been hesitant to invoke the miraculous in creation's history, might they consider that it would extend backwards to the early stages of the process by which hominins were being neurologically created in a manner that would someday enable them to image God?[71] Could it be that this is an example of what Amos Yong meant when he suggested that the laws of nature may function differently when the Spirit is working "from the future"? Or for that matter, could it be what James K. A. Smith called the Spirit working in more intense ways to carry out the purposes of God? I think the answer to each of these

69. See Namba et al., "Human-Specific ARHGAP11," 867.

70. Polkinghorne, "Kenotic Creation and Divine Action," 105.

71. Polkinghorne himself may have been unlikely to consider this; he may have thought it creates problems about the presence of natural evil. See Ritchie, *Divine Action and the Human Mind*, 319.

questions is, perhaps. But it is also conceivable that this specific event and the set-up associated with it occurred within the normal day-to-day functioning of the laws of nature that became part of God's long-range providential plan to bring humankind into existence. We just do not know enough about any single event like this to be able to explain how it happened, and furthermore, in my opinion, it is both scientifically and theologically naïve to try.

Fergusson writes that "the awareness of providence must lack the fullness of complete vision; it is received through a glass darkly." Then he goes on to say that as a component of faith, a biblical view of providence "should prevent the confident interpretation of each event as an element of a single divine blueprint."[72] Polkinghorne makes much the same point: An itemization of how God acts in specific events, he writes, is not possible, because "the cloudiness of unpredictability veils us from being able to make such a [conclusion]."[73] Furthermore, there is, Fergusson asserts, even a surd (chaotic) element in creation against which God works and this dynamic also makes it difficult to second-guess the specific nature of God's activity in particular events.[74, 75]

At the end of the day, acknowledging providence is a faith commitment. True, it is grounded in our reasoned belief in the resurrection and our own personal experience of life in God, but the existence of providence cannot be converted into a testable scientific theory. The activity of the God of the Bible is too complex and too unpredictable to be tied up by the tightly formulated "equations" and constrained precepts that characterize the repeatability demands of scientific theories. "As the heavens are higher than the earth, so are my ways higher than your ways and my thoughts than your thoughts," we are told in Isaiah 55:9.[76]

So, the belief that the creation of our species has taken place through providence is a faith commitment, not a scientific theory. However, at the end of the day, when it comes to the origin of our species, the scientific data do not lead to a well-defined algorithm that provides an alternative to the Christian view of how a species like ours arose. Instead, there is a

72. Fergusson, *Providence of God*, 42.

73. Polkinghorne, "Kenotic Creation and Divine Action," 101.

74. Fergusson, *Providence of God*, 42 (parentheses added).

75. John Webster wrote similarly: "Providence is not asserted on the basis of the insignificance of evil but on the basis of the belief that God outbids any and all evil." Webster, "Providence," 148.

76. Although the context for this particular verse in Isaiah 55 is its description of the mercy and forgiveness of God, in the context of Scripture as a whole it remains apt in a broader context as well.

near-unanimous agreement that our species exists here on earth because of a huge number of lucky breaks. Believing this also requires a faith commitment: that nothing exists outside the physical realm.

As I see it, this leaves two possibilities:

- Possibility 1: Nothing exists beyond the physical reality dictated by the initial properties of the universe's origins. Of the billions of earthlike planets that likely exist in our universe and perhaps even billions of universes, we happen to be in one that through a whole series of lucky events has given rise to a species with a full theory of mind, and this in turn has enabled the origin of language, art, culture, ethics, science, and religion. *It takes faith to believe this, but it is not irrational.*
- Possibility 2: The initial properties of the universe's origin and all that has happened since, including the origin of our species, has taken place because of, and within, the providence of God. Our species has been created so that we might share in and experience the goodness, love, and joy that emanates from within our Creator's very being. As Webster puts it, "Providence is . . . the fulfillment of the ordered fellowship with God, which is the creature's perfected happiness."[77] *It takes faith to believe this, but it is not irrational.*

Recall the first concern I wanted to examine: Can we single out specific events in the history of the evolutionary origin of our species and declare that a particular event happened in a manner that supersedes natural laws? I think the answer is no, but not because providence never works that way. Rather it is because our perceptive ability to understand all the details of specific events that occurred in the distant past is limited by the fact that we can only "see through a glass darkly."

Animal Suffering

Now for my second concern: the existence of the vast amount of animal suffering and death that has characterized the history of life on earth and appears to have played a significant role in the origin of our own species. For example, consider the arrival of the asteroid sixty-five million years ago, as described above. Without that event, scientists say, we would not be here. We can try to deflect God's responsibility for this by suggesting that God did not direct the asteroid's arrival. Rather, we might say, God only *permitted* a natural course of events to take place. But that attempt to absolve God of

77. Webster, "Providence," 152.

any responsibility only raises a new question—this time about God's power. Would we not expect an omnipotent God of love to have prevented the terror by tweaking the asteroid's trajectory? If God had wanted to clear a set of ecological niches in order to usher in the age of mammals, would not a more subtle process have been capable of generating the same effect? Some, including Charles Darwin himself, consider this question to be of such critical significance that their inability to resolve it causes them to lose their faith in Possibility 2, above.[78]

It is partly because of questions like this that many scholars in the science/theology community, not willing to let go of providence, have concluded that divine activity only occurs within the bounds of the natural laws described by science. Although God, according to this view, sustains the laws of nature, God does not interfere with them. Thus, by revealing the laws of nature, scientific analysis has, in essence, simply discovered how *God* works. Bad things (like destructive asteroids) happen as a natural consequence of how God, in divine wisdom and love, ordained that the universe should function: God created a universe which, by design, is free to create itself. As Sarah Lane Ritchie has shown especially poignantly, however, there are significant problems with this view, the most important of which is that it is difficult to square with the miracle-working God of traditional theology.[79] So even though this divine-action model may successfully let God "off the hook" for the pain and suffering associated with the catastrophic physical events which led to our arrival as a species, it comes at a significant theological price. I, like Ritchie, think it takes us too far away from divine action as it is portrayed both in the Bible and in the Christian tradition.[80]

Actually, this problem of animal suffering and pain is not a new theological challenge that arises only when we consider creation of new species through the evolutionary process. Although it is true that the horrendous suffering associated with the asteroid arrival led to the age of mammals and eventually our own species, some perspective on this matter is needed. In the one million years preceding that terrible crash, more than fifty thousand generations[81] of animals would have come and gone. Trillions of animals died in those million years, and many would have suffered excruciating pain as their lives ended. We have long known that the anguish of suffering and

78. Falk, "Theological Challenges."

79. Ritchie, *Divine Action and the Human Mind.*

80. So much more could be written about this, but it would take us beyond the scope of this book. I highly recommend Fretheim, *Creation Untamed,* for a thought-provoking approach oriented towards a general audience.

81. For a species with a generation time of twenty years there are fifty thousand generations in one million years.

death is a part of life. What is different about the impact of the asteroid crash is that a single generation of animals suffered their chaotic demise all at one time. It is that scene of agony that justifiably makes us cringe and causes us to wonder how God could ever allow such suffering to happen as a way of bringing about a whole new array of beings. However, we must not be so naïve as to think that just because it happened suddenly to one generation that a new theological problem is brought to bear. The existence of pain and suffering is an old theological challenge, one that has concerned theologians for thousands of years. For whatever reason, God has created a world where suffering and death exist. True, Christ chose to enter into this world and endured the excruciating agony of an ignominious death. True also, good frequently comes out of even the most dire of circumstances. However, we must not romanticize or gloss over the pain and suffering that has characterized our world for as long as there have been animals able to sense it. It certainly seems that even the joy and beauty that sometimes follows pain and suffering cannot sweep that agony away—not all of it and not for all suffering creatures. The cumulative pain of death and suffering might seem to far outstrip any good that can come from it, and I think it is appropriate to ask the question: why all this death and suffering? Could there not have been a better way for this world to have functioned over these hundreds of millions of years of animal life?

One of the most humbling things to me about the Bible is that it does not shy away from this sort of question. Indeed, as John Walton and Tremper Longman point out, the whole book of Job is the story of God being placed on trial for the seemingly unjust suffering in God's world. Intriguingly, the book of Job does not attempt to provide a pat answer. "The book of Job," they write, "turns our attention away from the idea that the world runs by God's justice and offers the alternative that, instead of trying to understand everything that happens as a reflection of God's justice, we should learn to trust [God's] wisdom."[82] We just do not know enough, and that is where God chooses to leave us.

We cannot and must not downplay the agony of suffering, but nor are we in a position to minimize God's wisdom and providence. As John Webster writes, "providence is not asserted on the basis of the insignificance of evil but on the basis of the belief that God outbids any and all evil. What makes evil problematic for providence is not its existence but the fact that we resist applying belief in providence to cases of it."[83]

David Fergusson is even more frank:

82. Walton and Longman, *How to Read Job*, 19.

83. Webster, "Providence," 148.

> The vexed relationship of the theology of providence to [the problem of evil] is unavoidable. Questions of suffering and evil now haunt the margins of every paragraph written on divine providence. But it is a mistake to assume that the function of providentialism is to provide the solution to a problem or that a successful [doctrine on the problem of evil] must be one of its essential ingredients. Once again, the history of theology is littered with flawed attempts to explain evil and suffering as encompassed within the providential purposes of God.[84]

What is important is to stay faithful to the Bible's message of trusting in God's wisdom and not to put God on trial again for a world in which suffering abounds. Job has already done this, and God has essentially answered: "Trust my wisdom, not your own."[85] As Webster writes,

> This is not a matter of theology evading responsibility to its setting but exercising it in the only way it knows. The task of a Christian theology of providence can be undertaken only by drawing on the resources given to it by the gospel; it can only hear the prophets and apostles, and only speak after such hearing—otherwise it has nothing to say. What it is not permitted to do is to respond to its contrary situation by suspending its talk of providence until better instructed, until better warrants are provided to ensure a more fluent, less crisis-laden exposition.[86]

So even though we are unable to solve this mystery as to why the God of love and harmony allows death and suffering, we are left with a choice, and it is the same choice with which God left Job. We are not God, and we do not have access to the full wisdom of God. Besides the book of Job, there are many other Scripture texts that wrestle with the reality of God's

84. Fergusson, *Providence of God,* 303.

85. This statement "Trust my wisdom, not your own," is nuanced in a manner that cannot be fully developed here. God does, after all, tell Job a bit about the nature of that divine wisdom. It *is* worthy of trust. Most of the terrible things that happened to Job occur because of natural evil, not moral evil. Job bitterly laments that God has created a world with so much chaos and he wants to know why doesn't God choose to govern the world in a more equitable fashion. In his profound analysis of how God answers Job, Fretheim writes: "To have elements of such disorder and insecurity is precisely the kind of world God wanted, for God did not intend the world to be a machine. God challenges Job so that he realizes the proper nature of that creation, that suffering may be experienced in just such a world, and so that he can learn what his place within it properly is, even in the midst of suffering." (Fretheim, *God and the World,* 310.) Fretheim's discussion of Job (pages 290 to 321 of *God and the World*) is well worth reading for those interested in exploring the nuances further.

86. Webster, "Providence," 149.

providence: cases where the biblical author seems to think God is working in a manner differently than expected. One of my personal favorites is the book of Habakkuk. After a "Job-like" conversation with God, Habakkuk ends with the matter still largely unresolved. However, like Job, his choice is to trust in God's wisdom, a choice not based on blind trust, but one which comes after careful reflection on who God is and what God has done. Here is one of my favorite passages in the entire Bible:

> I heard and my heart pounded, my lips quivered at the sound. Decay crept into my bones and my legs trembled. Yet I will wait patiently for the day of calamity to come on the nation invading us. Though the fig tree does not bud and there are no grapes on the vines, though the olive crop fails and the fields produce no food, though there are no sheep in the pen and no cattle in the stalls, yet I will rejoice in the Lord, I will be joyful in God my Savior. The Sovereign Lord is my strength; he makes my feet like the feet of a deer, he enables me to tread on to the heights.[87]

Again, we are not God. One of our greatest temptations is to be autonomous—to put our wisdom on a pedestal in place of God's.[88] This is so crucial a matter that both the Old and New Testaments begin by addressing it, but in different ways. In the garden, the crafty serpent is shown tempting Adam and Eve by telling them that if they eat of the fruit of the Tree of the Knowledge of Good and Evil, they could be like God. There was a limit to what they in their humanness were intended to know and the same applies to us. The temptation to think we can be autonomous is the most significant impediment to progress that our species will ever face. Intriguingly, the Gospel of Matthew also begins by describing a set of temptations. This time, it is Jesus who is in the docket. As we examine Satan's efforts to tempt Jesus, notice that the temptations are just the reverse of those to which Adam and Eve were subjected. Two of them begin with the words: "If you are the Son of God." Being fully human, Jesus is tempted to doubt his messianic role as the Son of God. Jesus, unlike Adam and Eve, dismisses them one by one. After this, we are told, he is attended to by angels. Notably, this is the converse of the angels with flaming swords which came to the garden in order to block Adam and Eve (with their humanized knowledge of good and evil) from accessing the Tree of Life.

Sorting out this matter of who is God, and who is not, is likely the single most important question our species will ever face. Choosing rightly is the most important answer.

87. Hab 3:16–19.

88. Bonhoeffer, *Creation and Fall*, loc. 1130.

CONCLUDING THOUGHTS

I have emphasized that there are likely many different ways in which God, through providence, has worked to accomplish the creation of our species. The process as a whole, we must not forget, emanates from God's desire that we experience life in the fullness of Jesus. With this in mind, I want to extend this discussion of providence to examine how it is still shaping us today. First, consider the church. There is coherence between the features of providence that we've been exploring in studying the origin of our species and those involved in bringing humankind towards maturity through the church. The church's founding principle is succinctly expressed in a single prayer—the conversation between Jesus and the Father in John 17. Just a few hours before his arrest, Jesus prayed, "that they be one [Father] as we are one—I in them and you in me—so that they may be brought to complete unity . . ."[89] This prayer for the church—that the many members might function as one—was taken up further by Paul as he wrote to the church in Corinth:

> Just as a body, though one, has many parts, but all its many parts form one body, so it is with Christ. For we were all baptized by one Spirit so as to form one body—whether Jews or Gentiles, slave or free—and we were all given the one Spirit to drink.[90]

The disposition that we said has so characterized the history of life on earth—the spirit of "better together"—has served as the guiding light for the long and winding road leading not only to our species but to all of life through all of time. Jesus' prayer for the church is consistent with this defining element of life itself.

In that prayer, Jesus also focused on the centrality of love as the guiding principle for the future church: "I have made you known to them, and will continue to make you known in order that the love you have for me may be in them and that I myself may be in them."[91] The consistency here with the reality of what we now realize is how our species came to be is profound. Similarly, Paul followed that great twelfth chapter of 1 Corinthians on how the church must function like a single body with his remarkable thirteenth chapter—the one on love.

Although these high and lofty ideals for the church's formation and existence have often not been realized, still—and this is the point I wish to emphasize—God's Spirit has been at work building God's church. In the

89. John 17:22b–23.

90. 1 Cor 12:12–13.

91. John 17:26.

messiness of life, in the reality of a world where evil still exists, there has been a progression, a long, two-thousand-year successive series of events through which the Spirit has been at work, building the church from that one little group of 120 up to where we are today. Every Sunday morning around the globe hundreds of millions gather to worship God and celebrate God's love and their love for one another. The Spirit is alive among these people and the Spirit is slowly but surely shaping them into the communities they are meant to be. I know; I am one of those persons whose life has been transformed through God's love at work through the people who have patiently loved me and lovingly cared for my family.

Central to these communities is another example of God's providential activity. At the core of the church is the Bible. Often people all over the world are reading the very same Scriptures and listening to a message centered on that word from God. The long set of providential developments that led to the formation and adoption of the biblical canon bear considerable similarity to the principles we have been discussing. The sixty-six (by Protestant accounting) books are a diverse array of different literary genres, each written in response to all kinds of crises by different authors at different times. Nonetheless, they function as a unified whole—assembled and woven together through the providence of God into what is likely the most influential (and beautiful, for that matter) literature ever written. One theme is central: love for God and respect for neighbor. Micah 6:8 summarizes this well, "He has shown you, O mortal, what is good. And what does the Lord require of you? To act justly and to love mercy and to walk humbly with your God." Each of the authors had no idea they were contributing to a massive work like this. They were just writing something down for their community. In some cases, it was only a letter to a single individual. In those early days as it was being put together there was a cacophony of voices with a variety of agendas calling for influence in its assembly. In and through those distractions, the Holy Spirit was actively laboring to bring about scriptural testimony to the God who is Love. Daniel Castelo and Robert Wall have written profoundly about this:

> Our affirmation of Scripture's oneness is grounded in a pneumatology that considers Scripture's formation and its formative performances to be an achievement of the Holy Spirit. We believe the Spirit selected a diversity of authored texts and then sanctified the postbiblical phenomena that ordered these authored texts into a biblical canon, which the church then recognized as effective for mediating all of God's word to all of God's people for all time. We believe this same Spirit who led in the

> production of Scripture now also leads in its performances to cultivate the church's ongoing communion with its living Lord.[92]

What they are describing here is providence at work over hundreds of years to give the church a foundation that has remained fully and completely relevant for all cultures through all time.

The writer to the Hebrews reflects back in the eleventh chapter on all that has gone on before as the interaction between human faith in the midst of one crisis after another played out within the providence of God. Hebrews 11 freely acknowledges that the long line of God's faithful people had not yet received all that had been promised. Together, though, they served as a vast cloud of witnesses testifying to the providence of God working in all sorts of different ways through the ups and downs, and twists and turns, that life brought their way. The point of the chapter is that God's providence had been working through all of that. And so it is with us. We can reflect on the providence at work in the formation of the church and what is still taking place through the work of the Spirit. We can reflect on the assembly of Holy Scripture and how in God's providence it is still brought to life in our daily existence. Moreover, and of equal significance, we can reflect on the origin of our species and our unprecedented ability to know, understand, and love each other and our God. As we do so, we are assured that the providence of God has been at work not only in our origin as a species, but continues to shape us into the image of Christ today.

As Webster writes, "Providence is that work of divine love for temporal creatures, whereby God ordains and executes their fulfillment in fellowship with himself. God loves creatures so he orders their course to perfection."[93] In the words of N. T. Wright, God has always been bringing us to a "genuine humanness" that "bears and reflects his image."[94] And, from David Fergusson: "we must have been intended; hence we are inevitable in the course of evolution."[95]

But how does the message of our arrival as a species through the evolutionary process, providential though it has been, relate to the Genesis story? How does the Adam and Eve scenario fit in with the scientific narrative? And if we were created through the evolutionary process, is our species still evolving here on earth? Are we able to draw out any message about that from Scripture? These huge questions are central not just to the origin of our species, but to its ongoing fate on earth as well; we turn to these next.

92. Castelo and Wall, *Marks of Scripture*, 50.

93. Webster, "Providence," 148.

94. Wright, *Day the Revolution Began*. 416.

95. Fergusson, *Providence of God*, 188.

7

Alpha and Omega: The Beginning and the End

ALPHA: THE BEGINNING

Genesis gives us two complementary accounts of human creation, each written from a different perspective. In the first, told in Genesis 1–2:4a, God has made and ordered the entire cosmos—heaven and earth. It is all God's dwelling place—the temple of God—and humans have been made in God's image to be the royal representatives in this sacred place. In the second, Genesis 2:4b–Genesis 3, the setting is a garden, where God is depicted as forming a person from the dust of the ground and bringing the person to life by breathing into his nostrils the breath of life. The two accounts are widely considered by Old Testament scholars to have been purposely stitched together, each emphasizing different aspects of the humans' place in creation.

How do the truths of these two complementary biblical accounts of human origins relate to what has been discerned about the origin of humanity when viewed through the lens of scientific investigation? First, let's examine what scholars perceive to be the main theological precepts that are set forth in these chapters that serve as a prologue to the biblical message.

GENESIS 1

Theological Reflections

Humans, the first account tells us, are created in the image of God. The Old Testament scholar Peter Enns helps put this into the context in which this would have been understood by Israel:

> Many scholars draw a parallel between the image of God in Genesis and images of kings in the ancient world. Rulers could not be everywhere at once, and travel was slow. So, they would erect monuments or statues of themselves throughout their kingdoms. These "images" let everyone know that the king's rule extended wherever his image was found.
>
> Another kind of image in the ancient world is an idol, a physical object that represented the god in the temple. Idols were not considered gods themselves. They were statues that let you know the god was in some mysterious sense "present."[1]

So that's the backdrop, and Enns goes on to write, "Statues of kings and of gods help us understand what it means for humans to be made in God's image: humans are placed in God's kingdom as his *representatives*."[2] We need to let the radicalness of this last statement sink in! Humans were created to be physical representatives of God in the world. In contrast, the Mesopotamian view was that at best most human beings were merely the gods' slaves. The cosmos functioned for the gods and human beings were merely an afterthought.[3]

Contrast this low view of humans with Psalm 8. There the author looks up into the heavens and contemplates the greatness of the moon and the stars, likely fully aware that many of his contemporaries considered them gods. Dazzled by the nighttime spectacle, and reflecting on the fact that all of this has been made by God, he reflects on his own infant-like puniness: "What are humans that you are mindful of them?" Then, bathed in the realization of God's love, his spirit starts to soar: We are "a little lower than God"; we are "crowned with glory and honor"; we are "the rulers over the works of [God's] hands." The message of Genesis 1 would have been understood, many Old Testament scholars[4] tell us, as a temple narrative. The whole cosmos is the sacred dwelling place of God. Other cultures had buildings that served as God's temple, but Genesis 1 is laid out in a manner that its audience would have understood as being the formation and ordering of God's temple. J. Richard Middleton writes:

1. Enns, "What Does Image of God Mean?" Although this quotation is from a blog post of Peter Enns, please see Middleton, *Liberating Image,* chapter 3 for a wonderful, detailed analysis.

2. Enns, "What Does Image of God Mean?"

3. Provan, *Seriously Dangerous Religion,* 83.

4. Middleton, *Liberating Image*; Walton, *Lost World of Genesis One*; Provan, *Seriously Dangerous Religion*; McConville, *Being Human in God's World*; also see Moberly, *Old Testament Theology,* in particular his summary of the work of Jon D. Levenson's *Creation and the Persistence of Evil,* beginning at loc. 627.

> This picture of creation as a cosmic temple also suggests the appropriateness of humanity as God's image in the symbolic world of Genesis 1. For just as no pagan temple in the ancient Near East could be complete without the installation of the cult image of the deity to whom the temple was dedicated, so creation in Genesis 1 is not complete (or "very good") until God creates humanity on the sixth day as imago Dei, in order to represent and mediate the divine presence on earth.[5]

There is something else that is unique about the creation of humans as described in Genesis 1. Here is a sampling of the step-by-step acts of creation:

- "Let the land produce vegetation . . ." Verse 11.
- "Let there be lights . . . " Verse 14.
- "Let the waters team with living creatures . . ." Verse 20.
- "Let birds fly above the earth . . ." Verse 20.
- "Let the land produce living creatures . . ." Verse 24.
- "Let *us* make humans in our image, in our likeness . . ." Verse 26.

In each case, God calls the created components into being. This is followed by an affirmation such as "so God made . . ." and then an assessment: "And God saw that it was good." But the creation of humans is formulated in the plural tense ("let *us*"); it is understood to be a communal affair. Who are these others with whom God is speaking, and do scholars think there is a significant reason for the plural tense in the case of humans? In the words of the late Terence Fretheim:

> A remarkable majority of scholars understand this plural in terms of the divine council, the heavenly assembly that engages the deity and does God's bidding. In addressing the divine assembly, God chooses not to create humankind alone, but with the aid of those beings who are not God . . . Israel understands that God is by nature a social being, functioning within a divine community that is rich and complex. God is not in heaven alone and, even more, is engaged in a relationship of mutuality within that realm.[6]

There is not space here to go through other examples of where Old Testament texts refer to this "divine council"; it is certainly not a well-developed scriptural concept—but it exists in other contexts as well. What is

5. Middleton, *Liberating Image*, 87.

6. Fretheim, *Creation Untamed*, 28.

important for us here is that the creation of humans is placed into a communal context. Here is what Fretheim has to say about the significance of this:

> In other words, relationship is integral to the identity of God, prior to and independent of God's relationship to the world. So, when God decides to work communally in creation, God is being faithful to God's own self . . . aloneness is not characteristic of God, and hence the isolated human being would not truly be created in the divine image . . . Only the human being as social and relational to other human beings is truly correspondent to the sociality of God and what it means to be created in the image of God.[7]

Middleton notes the use of the plural in Genesis 1:26, as well as the general wording[8] of verses 26–28, and states that it "leads to the exegetical[9] conclusion that the *imago Dei* refers to humanity's office and role as God's earthly delegates, whose terrestrial task is analogous to that of the heavenly court."[10] In other words, the audience for whom this passage was first written would have understood that being created in God's image on earth mirrored, in a sense, that which already existed in the heavenly realm—the assemblage of "angelic courtiers and emissaries." Middleton ends his analysis of how Israelites would have understood all of this with these words: "Imaging God thus involves representing and perhaps extending in some way God's rule on earth through the ordinary *communal* practices of human sociocultural life."[11]

J. Gordon McConville writes that "the fissuring of human community," which is in many ways the recurring story of the Old Testament, provides a sense that humanity is living away from its point of origin. "In their longing for such relating [as it is in the heavenly realm], humans know something of their origin in God."[12] To be created in the image of God means that we were created to live together—in harmony. Furthermore, God implants a longing—grounded in that image-of-God origin—to return to our harmonious roots.[13] Middleton points out that although the phrase "created in the image of God" is only specifically mentioned five times in the Bible, the

7. Fretheim, *Creation Untamed*, 28

8. More specifically, Middleton uses the phrase "explicit syntax."

9. I.e., based on biblical interpretation.

10. Middleton, *Liberating Image*, 59.

11. Middleton, *Liberating Image*, 59 (emphasis added).

12. McConville, *Human in God's World*, 28.

13. Augustine words it like this: "You have made us and drawn us to yourself, and our heart is unquiet until it rests in you." Augustine, *Confessions*, loc. 705.

concept itself actually runs through the entire Bible. He likens the concept to an underground river running just below the surface of the entire biblical message. It bursts forth from time to time, but it is never gone. Created in the image of God, where we are meant to live together in harmony with God and with each other, is a principle that underlies the entire biblical story.[14] McConnville writes of the importance of this manner of thinking—godlikeness as a *communal* characteristic—and he stresses how important this is in our contemporary context. It becomes the ground for "resisting every kind of denial of God-given freedom as a *condition of humanity*, including slavery, economic exploitation, and the subjugation of women."[15]

One of the places where being created in God's image "bubbles forth" is in the Letter to the Colossians. Here Paul calls on the reader to "put on the new self, which is being renewed in the image of its Creator."[16] The entire set of surrounding verses are communal in their orientation. For example, in the next verse Paul goes on to say that in the image of our Creator, "there is no Gentile or Jew, circumcised or uncircumcised, barbarian, Scythian, slave or free, but Christ is all, and is in all." And in the previous two verses Paul tells the Colossians that they must cast aside any hindrances that negatively impact their ability to function well in community. So this is how the Bible, God's written word, describes the origin of our species in Genesis 1 and beyond.

Scientific Reflections in the Context of Theology

God's work is also revealed through scientific investigation. My colleague theologian Mark H. Mann writes that the "two books" theory

> affirms that God's self-revelation is given to us, albeit in different ways, in both Scripture and Creation (the two books) and that Christians need to "read" Scripture and Creation together in order to understand the fullness of God's Word and truth for us today . . . The scientific enterprise is in many ways *sacred work*, for it is the attempt to understand more fully the handiwork of God, and is in this way not unlike disciplined reading and discerning the Word of God in Holy Scripture.[17]

14. Middleton, "Human Distinctiveness and the Origin of Evil." Also, for more detail, see Middleton, "Genesis Creation Accounts," 27.

15. McConville, *Human in God's World*, 28.

16. Col 3:10.

17. Mann, "Church Fathers."

This is not to equate the reading of the two books nor to put the Bible at the same level of significance as scientific investigation. But scientific tools are a gift from God too, and science rightly understood will complement knowledge gained from the Bible. So, with that in mind, let's return to what scientific analysis reveals about the origin of our species.

As the members of our species came to have increasingly sophisticated, socially adept minds so that they could successfully "theorize" what was on the mind of others, they were able to establish a complex web of thoughts about all sorts of things. Indeed, as they reached that level of thinking, they would also have been more conscious of the ramifications of whatever God's Spirit would impress upon their minds. That is, if they chose to listen! If one or more of those small communities chose to live in a corporate manner attuned to the Spirit's leading, they almost certainly would thrive. As time went by, genetic variants and cultural practices that increased the responsiveness to the Spirit would likely—at least in one or more populations—eventually become more common simply because such a lifestyle leads to communities that are healthier emotionally, and that in turn would lead to more offspring.

Jürgen Moltmann writes that the Spirit of God is "hidden, secret, and silent." On an everyday basis the Spirit is perceived as an "inexpressible closeness" and "companionship." It is this Spirit, he says, which is "the power of creation and the wellspring of life."[18]

We cannot say what percentage of the little communities lived in a manner that was grounded in Spirit-led wisdom at any given time. What we can say, however, is that when they did, they blossomed, and it was this that brought into fruition that uniquely human quality that Scripture labels as "the image of God." We are different than animals, God tells us multiple times in Genesis, because we bear that image.[19] Similarly, the Harvard-based anthropologist Richard Wrangham declares, we are vastly different than our ape cousins when one compares our lack of aggressiveness toward others in our group. Wrangham summarizes a comparative study between our species and the chimpanzee:

> Chimpanzees are not like people. A day spent with wild chimpanzees gives you a good chance of seeing chases, sometimes some hitting, while hearing fearful screams. Every month, you are likely to see bloody wounds. The primatologists Martin Muller, Michael Wilson, and I quantified the difference between an ordinary group of chimpanzees and a particularly disturbed

18. Moltmann, *Spirit of Life*, loc. 286, 573.

19. Gen 1:26–27, Gen 5:1, and Gen 9:6.

> population of Australian Aboriginals who had recently stopped hunting and gathering. Among the Australians, social disintegration and alcohol were considered responsible for raising the likelihood of physical aggression to especially toxic levels. However, even in this comparison with an unusually violent group of humans, the chimpanzees were several hundred to a thousand times more aggressive. The difference in the frequency of fighting between humans and chimpanzees is enormous.[20]

That's fine for the chimpanzee, but you'll recall our other close cousin species, the bonobo, is vastly more laid back than the chimpanzee. Despite their milder disposition, Wrangham indicates that they are far more aggressive than humans. "Physical aggression in humans happens at less than one percent of the frequency among either of our closest ape relatives."[21]

In describing this level of aggression, Wrangham is referring to something that psychologists refer to as reactive aggression.[22] Reactive aggression constitutes a rapid response to an immediate threat that involves anger, fear, or both. It is part of the fight-or-flight response, mediated by the rapid release of adrenaline. This is the kind of aggression that is unplanned, a sort of spur-of-the-moment response to an aggravating situation. It is this that has changed dramatically, Wrangham states, in our lineage.

I am not proposing that all was perfect even in the best of those little communities. What I am proposing, though, is that like a well-functioning Christian community that many of us have experienced in our churches (see chapter 7 in my book *Coming to Peace with Science),* those communities that responded to the Spirit of God were the ones (on average) that prospered. Inherent in this proposal is that our species emerged through something that is called "multilevel selection." To illustrate the concept, I want to turn your thinking back to those chicken cages we discussed in the last chapter. Recall that when investigators repeatedly bred from the most productive cages, it led to hens that were not only endowed with higher egg-laying capacity. In addition, they were characterized by a temperament conducive to harmony. In the wild, selective breeding of this sort, where the focus is on the most reproductively successful interbreeding group among a set of other such groups, is called "between-group" selection. In the other experiment with those chicken cages, investigators sought to identify the most productive hen within single cages each generation and then bred from that "winner." The wild counterpart of this type of selection, where a population's

20. Wrangham, *Goodness Paradox,* 19.

21. Wrangham, *Goodness Paradox,* 20.

22. Berkowitz, *Aggression.*

characteristics have emerged because of competition between individuals in a single group, is called "within-group" selection. In the chicken experiment, recall that using this alternative—within-group selection criterion—the most productive individuals had a very nasty temperament.

Evolutionary biologist David Sloan Wilson has been a strong advocate for the concept of multilevel selection for many years. He proposes that, in the wild, species that exhibit a high degree of sociality emerge when a population is comprised of small groups where the most reproductively successful group "wins out" at the expense of groups that do not produce as many offspring. In such cases, the primary force driving evolutionary change is an increased ability to function well socially; this is "between-group" selection as defined above. If, as may often be the case depending upon the species and breeding pattern, there is little competition *between* groups then the evolutionary dynamic is influenced most strongly by the most reproductively successful *individuals* within each single group. This corresponds to the single-cage experiment identified above and is thereby identified as "within-group" selection. There is, it appears to me, a growing consensus that the emergence of our species has been strongly influenced by "between-group" selection. Indeed, this was likely the key factor, many investigators propose, driving the emergence of the domestication-like qualities characterizing *Homo sapiens*.[23]

When David Sloan Wilson gives public talks about evolution, he regularly poses a question to the audience. He asks them to call out a single word that comes to their mind as they think of a hypothetical morally perfect individual—someone who epitomizes goodness. He has a slide prepared with all the words that he expects, based on previous talks, to be called out. Sure enough, the list essentially always includes "loving, honest, brave, giving, unselfish, loyal" and so on. Then Wilson asks his audiences to call out words for the opposite disposition, someone who is "evil." Here again he has a slide already prepared; the audience list always includes words like "greedy, murderous, selfish, deceptive, manipulative, and uncaring." Everyone has a moral compass embedded in their minds, he tells them. They have an inner sense of good and evil and are all pretty much in agreement regarding what each entails. From there Wilson goes on to state that what emerges when selection is acting primarily "within-groups" are gene forms that lead to "evil" traits like those listed above, whereas when selection occurs

23. Although the existence of multilevel selection has been the source of much debate over the years, I think the way that I describe it here reflects the consensus that has been emerging. For more detail, see *This View of Life*, by Wilson; *Social Conquest of the Earth*, by Wilson; *Survival of the Friendliest*, by Hare and Wood; and *Goodness Paradox*, by Wrangham.

"between-groups," the emerging set of traits are associated with goodness. As Edward O. Wilson and David Sloan Wilson put it in a 2007 review: "Selfishness beats altruism within groups. Altruistic groups beat selfish groups. Everything else is commentary."[24]

The emergence of our species is one of the eight major transitions in the history of life described by Maynard Smith and Szathmary (see last chapter).[25] With this last transition, however, life itself was taken to a whole new level. As de Chardin put it, our species had come not just to "know" like animals, individuals had come "to know that they know" and even more importantly, they had come to know that others know too.[26] This attribute changed the mind from solitary minds found somewhat independently within single individuals to a collection of minds, a sort of super-power.

Recall that in each of the eight transitions, a living system was taken to a new and respectively higher level of cooperation. For example, the eukaryotic cell (defined as a cell with compartments like the nucleus) was formed by the cooperative fusion between two prokaryotic cells (simple cells with no compartments). The smaller of the two fusing cells became the new energy producing compartment (the mitochondrion) of the new cell produced through this synergetic partnership. For this higher-level cooperative unit to function efficiently, the mitochondria have always had to be kept in check: If they ever started to multiply too fast—hogging more than their share of the cell's nutrients in a sort of coup—the system would fail.[27] In other words, the lower-level units must not become overly self-serving. There always had to be a way of keeping them under control to ensure that each part functions for the good of the whole.

As a second example, consider the transition from single-celled organisms to larger organisms with multiple cell types arranged into tissues and organs. Here too, cooperation is key. If any single cell type becomes self-serving and starts to grow at the expense of the other cell types, the functioning of the whole organism is negatively impacted. Cancer is one such example of where this happens. In essence, the whole body is taken hostage

24. Wilson and Wilson, "Rethinking the Theoretical Foundation of Sociobiology."

25. Maynard Smith and Szathmary specifically referred to the origin of language as the eighth transition. I, like David Sloan Wilson, am defining it a little more broadly here. I am suggesting that the origin of our species, with its multiple unique properties, can be thought of as the eighth transition, but this difference is largely semantic.

26. This last phrase, "to know that others know too," did not come directly from de Chardin, but certainly the concept did. See de Chardin, *Phenomenon of Man*, 165–76, and *Hymn of the Universe*, 103.

27. For a recent paper exploring selfish mitochondrial DNA, see Gitschlag et al., "Nutrient Status Shapes Selfish Mitochondrial."

through having been sabotaged by one specific cell lineage. Through the eons of time our bodies have become equipped in various ways to keep this horrific possibility in check, but the frequency with which those defensive mechanisms fail testifies to the existence, on occasion, of renegades which selfishly seek to take over the whole body's supply of nutrients to care for their own needs. The fact that the grand parade of organisms on earth, in all its magnificent diversity, has continued for 3.85 billion years is a testament to the success of the cooperation strategy. The challenge, however, has always been to find ways of keeping any self-serving entities at bay: Individual units will take advantage of the cooperating whole if given the chance.

If this tendency for self-centered units to emerge is true for the other seven transitions, wouldn't it be an ongoing challenge to social cohesion (the eighth transition) as well? Wouldn't self-centered individuals continuously arise within groups taking advantages of the cohesion, thereby preventing it from emerging in the first place or, at the very least, present an ongoing challenge to group functionality? Here is what David Sloan Wilson has to say about this:

> We are evolution's most recent major transition. Almost everything that sets us apart from other primate species can be explained as forms of cooperation that evolved by between-group selection, thanks largely to our ability to suppress disruptive within-group selection. In most primate societies, group members are cooperative to a degree but are also riven by within-group conflict. Even the cooperation that exists often takes the form of coalitions warring with other coalitions within the same groups. To the best of our current knowledge, our distant ancestors evolved the ability to suppress bullying and other disruptive self-serving behaviors within groups, like multicellular organisms evolved ways to suppress cancer cells, so that the primary way to survive and reproduce was through teamwork.[28]

It is important to take special note of what Wilson writes about the problem of group cohesion in other primate species, where social structure only emerges to a limited degree. Selfishness keeps popping up, disrupting the cohesion of the group, he states. A new higher-level cooperative unit can only make it through a transition if there is a way of keeping selfishness of individual units in check. With that in mind, the nagging question is, what happened in the origin of our species to keep bullying in check? Wilson writes that it is a common currency of morality—an inward sense of good and bad behavior. There is a consensus about what the rules entail because

28. Wilson, *This View of Life*, 90.

they are an unwritten code ingrained in their minds. But, he says, even this will not work unless there is a sort of policing mechanism—some way of suppressing those who would seek to take advantage of the "goodness" of the cooperators. Wilson writes,

> Our moral psychology is the societal equivalent of cancer-suppressing mechanisms in multi-cellular organisms. The coercive side [that is, the "policing" mechanism] of morality is required to suppress the potential for disruptive self-seeking behaviors within groups. Once the coercive side is established, then it becomes safe for group members to freely help each other without fear of exploitation.[29]

The qualities of morality that Wilson describes is the equivalent to what Paul describes as he writes about the Gentiles:

> Indeed, when Gentiles, who do not have the law, do by nature things required by the law, they are a law for themselves, even though they do not have the law. They show that the requirements of the law are written on their hearts, their consciences also bearing witness, and their thoughts sometimes accusing them and at other times even defending them.[30]

As we think about deep history, what Wilson is describing is the same phenomenon that Paul describes; it is just deeper back in time. This "writing on the heart" served as a "policing" mechanism. It would not necessarily have required the capital punishment of bullies, as one investigator has proposed.[31] What was required to set this initially in place at the genetic level may well have been a garden-of-Eden-like experience for some period of time where individuals, under the guidance of the Spirit of God, were effectively policing themselves, holding each other accountable in a manner that was good for the group as a whole. This of course is the hypothesis central to the thesis of this book.

Wilson, however, does not think goodness is built into creation. He writes, "There is no warrant for a worldview that claims that nature writ large embodies goodness. Goodness only emerges when between-group selection pressures outweigh within-group selection pressures."[32] Even though Wilson shows great respect for religion and uses church small groups as the model for how society can begin to thrive, he acknowledges that his own

29. Wilson, *This View of Life*, 91.

30. Rom 2:14,15.

31. Wrangham, *Goodness Paradox*.

32. Wilson, *This View of Life*, 78.

personal beliefs are atheistic. With that, he cannot help but conclude that goodness is not an overarching principle inherent in the universe. No ideal, no value can be considered "writ large in nature" if all that happened is the result of one long series of accidents. However, there is solid reason for taking the opposite view: Goodness *is* built into creation[33] because the Creator is good—perfect, actually! It is the very essence of goodness—God's Spirit—that beckons individuals to take the right course of action. John Wesley, 250 years ago, referred to this ever-present prompting of the Spirit as prevenient grace. Michael Lodahl summarizes this well:

> For Wesley, there is no human being who has ever existed who was not deeply loved and immediately graced by God . . . Thus, no human being lives without the Spirit's wooing to live compassionately and justly with (all) others . . . "Everyone has some measure of that light, some faint glimmering ray, which sooner or later, more or less, enlightens every man that cometh into the world."[34]

So, the story of Genesis 1 is that the creation of humans living into their role as the Creator's faithful representatives on earth is what makes creation not just good, but makes it *very* good. There is, however, a complicating twist to human existence and to begin to understand that we need to turn to the second (and complementary) biblical account of our creation.

GENESIS 2

In both of the two creation accounts there is a focus on relationship. In the first, the story starts with the broader world and then ultimately moves to us showing how we are intended to fit into the whole. As each of the major steps of creation are completed, God assesses the outcome and declares it to be good. On the sixth day, the creation of the land animals is completed and the Lord's assessment, again, was that it was good. Then human beings were created, but this time there is no assessment of this specific creative act. Indeed, as Terence Fretheim has observed, human beings "are not even given a separate creation day but rather share the sixth day with the animals."[35] It is only after God looks out at all that God has made, including humans with their relational role in the whole creation, that God makes the assessment

33. I want to emphasize that I am not implying that humans are inherently good today. We will come to a discussion of original sin shortly.

34. Lodahl, *Matthew Matters*, 82.

35. Fretheim, *Creation Untamed*, 12.

that creation is "very good." The details of how our species was created or even any assessment of the quality of individual human attributes is not addressed. What is important in the Genesis 1 narrative is our ongoing role. We are the relational species. We have a unique relationship to God (as God's image in the world) and to all of creation (as its stewards).

The second account, which starts in verse 4b, begins with a dry and barren Earth. There had been no rain so there was no vegetation and furthermore, we are told, there was no one to till the ground anyway. As the story unfolds, we are told that there was moisture welling up from below the ground generating a mist, and it is at that point that God forms the first human[36] from that earth. One has the feeling in reading this account that God is palpably present. We can picture God bending down and forming a person from the soil, and then breathing into him the breath of life, enabling him to become a living being. So, in this account, the human is created first—not last—and human is singular, not plural as was the case in the first account. As mentioned above, most scholars today think that the two accounts were knit together on purpose and that they were meant to complement each other.

Next, we are told that God planted a garden that had beautiful trees bearing abundant fruit for nourishment, and it was there that God placed that first human. Two trees are singled out and they become especially important to the story. The first is called the Tree of Life and the second the Tree of the Knowledge of Good and Evil. Eating of the latter tree, the human was told, would result in death. Nothing more is said about the Tree of Life, although it becomes clear later that it is this tree and the human's access to it that the story is really all about.

We are told there were four rivers that watered the garden, two of which would have been well known to the audience, but the other two, although named, would have been mysterious. In examining this and other geographical aspects of the story, Old Testament scholar Iain Provan writes that "the authors of Genesis almost certainly did not have a particular location in mind when writing about this garden." The garden described in Genesis 2, he suggests, was

> not located somewhere *on* the earth but represents the whole earth. It is the very place where we find all the plants and trees, all the beasts and birds and all the images of God (human beings) who are placed in God's temple-cosmos. Put succinctly, the

36. The first human is not given a name at this point. The Hebrew word, 'adam,' is a generic word for humankind. The first time the person is actually named—Adam—is in Genesis 5:1.

> creation narrative of Genesis 1 is retold in Genesis 2, this time through the metaphor of the garden rather than the temple.[37]

Once the newly created human is placed in the garden of Eden, the Lord God makes an assessment, this time declaring the very opposite of what we saw at each step in the first chapter. This time creation contains a "not good" element; specifically, we are told that God said, "it is *not good* for the man to be alone."[38] Provan notes this and writes, "It seems that the human vocation in the cosmos can be fulfilled only in community, not individually."[39] Fretheim similarly evaluates this negative assessment that God makes as he reflects on who God seems to be in conversation with:

> To whom is God speaking here? It is likely that the reader is once again permitted to overhear the inner-divine reflective process, providing a connection back to the "let us" of 1:26; the 'a-da-m's not being alone is correlated with God's not being alone. Here God identifies a problem within the divine creative work up to this point and moves to find a solution, to make those changes that would enable a different evaluation: it is now good.[40]

So what does God do to move towards that different evaluation? As J. Richard Middleton notes:

> The first step towards resolution occurs when YHWH God forms animals from the ground (from which the human was also formed) and brings them to the human "to see what he would name them" (Gen 2:19). The fact that YHWH God does not bring animals to the human *to see if any of them would be a suitable helper* suggests that this is not trial-and-error on the Creator's part . . . YHWH God is aware that the animals will not fulfill the basic human need specified in Genesis 2:18. But it is not enough for the Creator to know this; the *human* needs to recognize this for himself. The text thus suggests that humans need to be active participants in our own flourishing (even though such flourishing is ultimately a gift from beyond ourselves)—such is the dignity of the creature in relation to our Creator.[41]

37. Provan, *Seriously Dangerous Religion*, 42.

38. Gen 2:18.

39. Provan, *Seriously Dangerous Religion*, 96.

40. Fretheim, *God and the World in the Old Testament*, 92.

41. Middleton, "Primal Harmony," 153.

Middleton's point needs to be emphasized. Too often we have an authoritarian view of the Creator as we think about human creation. But as Middleton and these other scholars point out, this is not an accurate biblical perspective. Rather, in the Bible as a whole, God is shown to be constantly encouraging human beings to participate in the shaping of what they will become as individuals and as a community. That divine-human interaction starts here in the creation story, but it is a constant theme throughout the entire biblical narrative.[42]

Let's pause in this discussion of the biblical narrative to reflect on what we discussed earlier based largely on the scientific analysis of how our species came to be. Recall Darwin's conjecture (page 15) 150 years ago that if our human ancestors were not "ferocious," then there would have been a distinct selective advantage to becoming increasingly cooperative over time. Remarkably, Darwin's speculation comes close to how many scholars of human evolution now view the secret to the success of our species. One of those is Harvard's Wrangham, who wrote:

> Given how much less emotionally reactive humans are than chimpanzees, bonobos, or most group-living primates, a low propensity for reactive aggression cannot be taken for granted . . . Reduced reactive aggression must feature alongside intelligence, cooperation, and social learning as a key contributor to the emergence and success of our species. Docility should be considered as foundational of humankind, not just because it is unusual, but because it seems likely to be a vital precondition for advanced cooperation and social learning.[43]

The question is, how did we embark upon this road to docility? Most of the scholars in converging disciplines of paleoanthropology, archaeology, paleogenetics, primatology, and evolutionary psychology are seeking to explain it on the assumption that there is nothing outside of physical reality influencing the process. However, the thesis of this book is that there is a different explanation—every bit as rational. It is laid out for us in broad form in the early chapters of Genesis. Too often Christians have thought that our creation came about only by divine decree alone. But that is not the

42. Recall the biblical passage to which we referred earlier: "The word is very near to you; it is in your mouth and in your heart for you to observe. See, I have set before you today life and prosperity, death and adversity . . . I call heaven and earth to witness against you today that I have set before you life and death, blessings and curses. Choose life so that you and your descendants may live, loving the Lord your God, obeying him, and holding fast to him; for that means life to you and length of days" (Deut 30:14–15, 19–20a).

43. Wrangham, *Goodness Paradox*, 124.

picture we get from the Bible. Rather, the most parsimonious biblical and scientific explanation of how our species came to acquire these key qualities is that they emerged from within a divine/human partnership.

Having stressed this observation, let's get back to the Genesis 2 narrative. Provan points out the significance of the human assigning names to each of the animals. "Their names," he says, "speak not so much of what they *are* as of where they fit in the cosmos." In Genesis 1, it is the act of God speaking that helps to order the world and push back chaos. Now in Genesis 2, it is the human, acting at God's behest (and in the image of God), that accomplishes this.[44]

Middleton makes the same point in comparing the first account to the second:

> [In Genesis 1,] while God names the various realms or spaces that have been separated, God does not name any of the inhabitants of those realms. Why does God refrain from naming these creatures? Perhaps because the creator does not want to hoard this prerogative, but, on the contrary, wishes to give space for humanity to complete this privileged task. However, humans do not name anything in Genesis 1. We have to wait for Genesis 2, where the first human indeed names the animals, as an expression of human rule as the image of God.[45]

So how ought we think about this peculiar passage in Genesis 2 where God is shown bringing each of the newly created animals to the human for naming? Whatever way it would have been understood in detail, one thing is for sure—the human is co-participating with God in the creative process. This, in contrast to the neighboring cultures of the day, would have been astonishing.

At any rate, after all that, creation was still "not good." "But for the man," the narrator tells us, "there was not found a helper as his partner."[46] Creation was not complete. As Iain Provan observes, "Only in community can humanity fulfill the human vocation."[47] The chapter concludes with God creating the woman and bringing her to the man. With that, creation could be considered complete and this time it is the human—not God—who assesses the result. "Bone of my bone and flesh of my flesh," he declares. It is only as we recognize and celebrate the significance of the other—with that mind much like our own—that we become truly human.

44. Provan, *Seriously Dangerous Religion,* 97.

45. Middleton, *Liberating Image,* 293.

46. Genesis 2:20b, NRSV.

47. Provan, *Seriously Dangerous Religion,* 98.

That is the biblical message about the origin of our species, and it ends with one final intriguing observation: "And the man and his wife were naked and were not ashamed."[48] Why would a statement like this serve as the capstone of human creation? As Kathleen M. O'Connor has written:

> The two figures are unashamed of their bodies. They are whole and integrated in themselves. They live in complete openness before one another like innocent children. Nothing is amiss within or without in this Garden. Comfortable with and accepting of themselves as embodied beings, they are also at peace with their surroundings, with the animals and with God.[49]

GENESIS 3

As attention-getting as this "naked and unashamed" statement is, the narrator does not let the matter rest there. In the very next verse, a new character enters the story: the serpent. This, we are told, is the craftiest (or most cunning) animal that the Lord God had made. As the story unfolds, we find that it is his craftiness—the sneaky reasoning he initiates—that leads to Adam and Eve's loss of innocence. The Hebrew word for crafty is *arum*, and the word which the author uses for nakedness is *arummim*—a play on words, used for emphasis. The word for nakedness appears once again just a few verses later, this time after Adam and Eve have disobeyed God. Now we are told all innocence associated with the naked state is gone; they feel ashamed and need to cover up. This is not the place to tease apart the details of the extensive Hebrew wordplay around the use of *arum* and *arummim*. It is addressed more fully by O'Connor,[50] Provan,[51] and especially in a study by Ellen van Wolde[52] on the semiotics of Genesis 2 and 3. In essence, the narrator sets a pivot (or fulcrum) centered on the serpent's craftiness. On either side of the pivot is Adam and Eve's mental state. Prior to being drawn into a dialog with the wily serpent, they were naked and unashamed—like young children in their innocence. On the other side of the dialog, following their disobedience and no longer innocent, their nakedness was not only a source of shame, but it was even accompanied by a sense of fear in the presence of their Creator. They knew that all was not right between them and their God, and they tried—somewhat inadequately apparently—to make fig

48. Gen 2:25 NRSV.
49. O'Connor, "Why Read the Story in the Garden," loc. 567.
50. O'Connor, "Why Read the Story in the Garden," loc. 589.
51. Provan, *Seriously Dangerous Religion*, 118.
52. Van Wolde, *A Semiotic Analysis of Genesis 2–3*, 165.

leaves function as loin cloths. The innocence and the beauty of their lives were gone. Whereas earlier Adam had joyfully affirmed his oneness with the woman ("bone of my bone and flesh of my flesh"), now he blames her for his sin: "The woman you put here with me—she gave me some fruit and I ate it." He seems to be shifting blame, first on God for giving him such an inadequate partner, and then on the woman upon whom he had earlier heaped praise. Adam is now focused on himself: alone again.

The Scripture is quick to point out that God does not abandon Adam and Eve in this lost and forlorn condition. Provan summarizes God's response:

> Into this broken situation, however, steps God, who is intent on providing a more lasting solution to the problem. He makes "garments" for them (3:21)—a word used elsewhere in the Old Testament for the main ordinary garment of a person in ancient Palestine (a long, knee-length or ankle-length shirt). If there must be clothing, at least it shall be proper clothing. God has essentially accommodated himself to the new reality of shame in the world, and he has become actively involved in finding the best way forward, given all the circumstances. From the perspective of the biblical authors, God remains the God who is "for us," even in the midst of our wrongdoing and shame.[53]

The primal sin, Fretheim emphasizes, is the "mistrust of God and God's word."[54] In his masterful book *Creation and Fall*, Dietrich Bonhoeffer describes where this mistrust took Adam and Eve and where it takes us today. In that initial dialog, the serpent tells Eve that by eating the fruit she can become "like God" (*sicut Deus* in Latin). Although the conversation is depicted as being between the serpent and Eve, it is also apparent that Adam was there; he was a silent and affirming "partner-in-crime"; they were in this together. The disobedience seems unfathomably ill-conceived. Adam, after all, had been given the important functional task of naming the animals as God brought them to him one at a time, and moreover, we are told in Genesis 1, Adam and Eve were God's royal representatives; it says they were created in the image and likeness of God on earth. The serpent told them, however, that they could have more—that they could be "like God"—and all they would need to make this possible was to eat the forbidden fruit. The choice was made and here is how Bonhoeffer profoundly describes its ramifications:

53. Provan, *Seriously Dangerous Religion*, 259.

54. Fretheim, *God and the World in the Old Testament*, 113.

> *Imago Dei*—humankind in the image of God in being for God and the neighbor, in its original creatureliness and limitedness;
> *Sicut Deus*—humankind like God in knowing out of its own self about good and evil, in having no limit and acting out of its own resources, in its aseity, in its being alone.
>
> *Imago Dei*—bound to the word of the Creator and deriving life from the Creator;
> *Sicut Deus*—bound to the depths of its own knowledge of God, of good and evil.
>
> *Imago Dei*—the creature living in the unity of obedience;
> *Sicut Deus*—the creator-human being who lives on the basis of the divide between good and evil.[55]

The story is heartbreakingly sad: Adam and Eve allowed themselves to be seduced into exchanging their rightful place before God by putting themselves in the place intended for God. But as scandalous as this seems to have been, the fact is that it is our story too. We also choose to remove God from the pedestal on which our lives ought to be based, and far too often we choose to build our own foundations, living in our own self-focused manner instead of God's. But Bonhoeffer goes on and reminds us that when that protective covering that we sew together for ourselves starts to unravel, God has not left us comfortless:

> *Imago Dei, sicut Deus, agnus Dei* [Lamb of God]—the human being who is God incarnate, who was sacrificed for humankind *sicut Deus*, in true divinity slaying its false divinity and restoring the *imago Dei*.[56]

I am writing this a few days before Pentecost Sunday, the day in the church year that celebrates the coming of the Holy Spirit into the lives of the apostles. After Christ had ascended, his followers had been told to wait for the coming of the Spirit's empowering presence in their lives. So they waited, and what happened next changed the course of history. Indeed, by Christ and through the Spirit, their "false divinity" had truly been slain and they were living out their vocational calling. *Imago Dei*, indeed.

Christians today are called to live into the future, not in their own strength, but by the power of the Spirit of God. We are called to be new

55. Bonhoeffer, *Creation and Fall*, loc. 1179.

56. Bonhoeffer, *Creation and Fall*, loc. 1179.

creations in Christ. The *Imago Dei* in each of us can be restored and we can begin to fulfill the purpose for which we have been created.[57]

Bonhoeffer makes this rhetorical jump from the garden of Eden into the garden of Gethsemane and onto Calvary to make sure that the reader fully grasps that there is a solution to the problem of the fall. Having made that point, his little masterpiece quickly returns to the Genesis story, which we will now do as well.

THE PROPAGATION OF SIN: GENESIS 4 AND BEYOND

The biblical narrative emphasizes the contagious implications of Adam and Eve's sin. Sin lingers, debilitating the human condition in a manner that spreads through the generations like a virus that won't go away. Genesis 4 begins by showing us how the jealousy of Cain seeds his own inner destruction. But it also shows us how God—far from abandoning humankind—is still there, seeking to bring out the best in Cain. Clearly, God has not given up on humanity and God gently admonishes Cain as he flirts with evil:

> "Why are you angry?"
> "Why is your face downcast?
> "If you do what is right, won't you be accepted?"
> "But if you do not do what is right, sin is crouching at your door;
> it desires to have you, but you must rule over it." [58]

Can you imagine a better description of how easily bitterness can well up within our own hearts? Intriguingly, God simply warns Cain; the choice is his. God's beckoning presence is never far away, but Cain must choose.[59] The next thing we know is that he has chosen to live his way and not God's, as he invites his brother to take a walk with him to a field. There, Cain

57. It is tempting to be disillusioned by the many *imago Dei* misrepresentations that characterize certain congregations, pastor-leaders, political figures, and ordinary individuals who self-identify as Christians. Jesus said, "Many will say to me on that day, 'Lord, Lord, did we not prophesy in your name?' . . . Then I will tell them plainly, 'I never knew you. Away from me, you evildoers!'" (Matt 7:22–23). And the apostle Paul, faced with all the challenges of building the early church, wrote: "For the time will come when people will not put up with sound doctrine. Instead, to suit their own desires, they will gather around them a great number of teachers to say what their itching ears want to hear" (2 Tim 4:3). The hypocrisy of individuals, or groups who claim to be followers of Jesus, must not be a detractor from our own personal responsibility to live into that for which we have been created.

58. Gen 4:6–8.

59. The profundity of this choice is the centerpiece of John Steinbeck's novel *East of Eden*, widely regarded as one of the great novels of American literature.

attacks and kills Abel. Of course, Cain must endure the consequences of his sin, and God tells him that he will now be a lonely wanderer. Once more, the sinful human feels miserably alone. Cain becomes fearful, just like his parents after their sin. From within his guilty state, he agonizes that anyone who finds him will kill him. But God, sensing his anguish, tells Cain he will be protected as he puts a mark on him to ensure his safety—an echo, perhaps, of the garments God made for Adam and Eve to cover their shame-filled existence as well.

As the fourth chapter continues, we see the human predicament spiraling downward. Lamech, one of Cain's descendants, also commits murder. Haughtily emphasizing how much worse he is than his forefather, he brags: "If Cain is avenged seven times, then Lamech seventy times seven."[60] Sin is spreading and getting worse and worse as the fourth chapter comes to an end. But the debasement of the human condition continues its downward descent, and in chapter 6, God, we are told, has a "deeply troubled heart" and regrets the decision to create humans: "How great the wickedness of the human race had become on the earth and that every inclination of the thoughts of the human heart was only evil all the time."[61] Rather than give up, however, God chooses to start again; God restores order and does so in a manner that harkens back to the original creation out of the watery chaos described in Genesis 1:2. Life begins all over again. Indeed, we are reminded in Genesis 9:6 that humans have been created in the image of God and that Noah, reminiscent of how Adam was formed, is "a man of the soil." Noah plants a vineyard—a new garden for a new day. Alas, however, we quickly learn that all did not stay well in that garden either, as the narrator tells us of a drunken Noah laying naked in his tent. Shamefully, one of his sons sees him in his nakedness. His other two sons are able to drape a garment over his body without seeing him by backing into the tent, but the damage has been done: Noah's naked and shameful state has been exposed. He wakes up angry as can be at having been seen in his naked state, and proceeds to punish his son Ham's offspring, even though Ham may have done nothing worse than look at Noah in his shame-filled condition.

As the sad story continues, we arrive at Genesis 11 where we are shown humans in their "*Sicut Deus*" condition building a tower to the heavens in order to make a name for themselves. Provan helps puts this story into perspective:

> For a mortal being to make a name for himself, then, is inappropriate. It is to seek to be "like God" in a wrong way—the

60. Gen 4:24.

61. Gen 6:5.

> "original sin" of Genesis 3. This is the significance of the Tower of Babel. Its builders want to make a name for themselves, and they are prepared to disobey God to achieve this "heroic" goal. Of course, their attempt at human solidarity in opposition to God, co-opting religion in the service of self-worship, is inevitably doomed, precisely because they are not God.[62]

Again, we can see that God has still not given up on humankind despite their sinful disposition, for the very next thing we are shown is God beginning to work with Abraham, through whom "all peoples on earth will be blessed" (Gen 12:3). For the tiny little band of individuals who passed on this story of their heritage from one generation to the next in those early days, little could they have imagined just how true this promise would become. The God who made that promise to Abraham thousands of years ago has not given up on humankind. Thousands of years later, it has become the historical foundation of the faith journey for billions of us.

As Bonhoeffer concludes his *Creation and Fall*, he takes the reader back to that Tree of Life standing in the middle of the garden. He writes that in a sense the tree reemerges in the New Testament. It is on a hill called Calvary. Of all the words I have ever read about creation, I think the most profound are the following words from Bonhoeffer. Humans freely chose to set *themselves* up as gods, pushing God and God's wisdom off to the side. That, far too often, is still true today. However, the Tree of Life remains in place and each of us, depending on our choice, would be welcomed to live in its presence:

> The end of Cain's history, and so the end of all history, is Christ on the cross, the murdered Son of God. That is the last desperate assault on the gate of paradise. And under the whirling sword, under the cross, the human race [serving as its own god] dies. But Christ lives. The trunk of the cross becomes the wood of life and now in the midst of the world, on the accursed ground itself, life is raised up anew. In the center of the world, from the wood of the cross, the fountain of life springs up. All who thirst for life are called to drink from this water, and whoever has eaten from the wood of this life shall never again hunger and thirst. What a strange paradise is this hill of Golgotha, this cross, this blood, this broken body. What a strange tree of life, this trunk on which the very God had to suffer and die. Yet it is the very kingdom of life and of the resurrection, which by grace God grants us again. It is the gate of imperishable hope now opened, the gate of waiting and of patience. The tree of life,

62. Provan, *Seriously Dangerous Religion*, 179.

> the cross of Christ, the center of God's world that is fallen but upheld and preserved—that is what the end of the story about paradise is for us.[63]

CONGRUENCY

So that is the biblical story of the origin of our species and what the Bible has to say about our current circumstance. Let's return now to the scientific story on which so much of this book has focused. How does the scientific story compare to all of this? The fact is that it is profoundly congruent. The human lineage has become highly distinctive in its ability to overcome the reactive aggression that, by comparison, has restricted the sociality of other primates. Wrangham's statement, discussed above, is worth repeating: "Docility should be considered as foundational of humankind, not just because it is unusual, but because it seems likely to be a vital precondition for advanced cooperation and social learning."[64] That sounds similar to me to an "Eden-like" existence—a small population living in relative harmony with each other and the Spirit of God. We know from the experiences documented throughout the preceding chapters that when animals are purposely made less aggressive through selective breeding, their genetic make-up is altered. The animals become less aggressive, in other words, because certain genes have changed—and to some extent we know which genes and we are learning how they have changed. With regard to our own lineage, many of those same genetic changes have been occurring and, undoubtedly, they played no small role in our success as a species. Something happened to bring this about, and as discussed in the preceding pages, there is much material in the literature about how this could have happened. There is even a term for it that is now widely used—self-domestication. What the secular scholars don't discuss is the possibility that the community (or communities) may have been living in this "docile" manner because they had chosen to live in harmony with the Spirit of God. Recall once more what Paul said about gentiles who didn't have the law:

> When Gentiles, who do not possess the law, do instinctively what the law requires, these, though not having the law, are a law to themselves. They show that what the law requires is written on their hearts, to which their own conscience also bears witness; and their conflicting thoughts will accuse or perhaps

63. Bonhoeffer, *Creation and Fall*, loc. 1496.

64. Wrangham, *Goodness Paradox*, 124.

> excuse them on the day when, according to my gospel, God, through Jesus Christ, will judge the secret thoughts of all.[65]

If this "writing-on-the heart" applies to the Gentiles who did not formally "possess the law," was there a specific time a few thousand years ago when this "writing-on-the-heart" process began to be true? Or alternatively, could it be that this process stretches far back in time—all the way back to the early days of our species? In the primeval biblical history of Genesis 4, how did Cain know it was wrong to murder his brother? There was no written law yet. Was it not because it was "written on his heart" by the Spirit of God? Where did Noah's sense of shame come from in that episode with his sons? There was no written law yet—but clearly there was a primal, heartfelt sense of what is right, and Noah and his sons knew that it had been violated. These "writing-on-the-heart" examples occurred in primeval history and they show that long before the existence of the nation of Israel, God's Spirit was at work guiding people into the recognition of right and wrong behavior.[66]

The scientific story does not stop with the formation of a remarkably cooperative species in which individuals live in docility in what "seems likely to be a vital precondition for advanced cooperation and social learning."[67] Although it is true that probably no other primate species comes close to the calm demeanor associated with our low *reactive* aggression, our level of *proactive* aggression—the kind of aggression that plots out how to advance one's own cause at the expense of others—is not low.[68] Far too often we plot out how to get even with those who we perceive as doing us harm. What makes our situation as a species unique is that we are wholly conscious individuals with an unprecedented ability to think about the mindset of others. This opens the door to a unique sort of craftiness, one that puts us in a position to knowingly choose between good and evil. Given the relative freedom that characterizes human existence and our ability to craftily plot schemes for our own advancement, the lure of self-centeredness and its subsequent spread has always been a distinct possibility. We might say that even from the scientific perspective it has always been "crouching at the

65. Rom 2:14–15.

66. For a thorough analysis and outstanding discussion of this point, see Fretheim, *God and the World in the Old Testament*, especially the fourth chapter, where he examines God's involvement with Israel's neighbors as discussed in Genesis 12–50.

67. Wrangham, *Goodness Paradox*, 124.

68. As Wrangham puts it, "We are low on the scale of one type (reactive aggression), and high on the other (proactive aggression). Reactive aggression is the "hot" type, such as losing one's temper and lashing out. Proactive aggression is "cold, planned and deliberate." (Wrangham, *Goodness Paradox*, 9.)

door" of human reality.[69] But ought we expect that there was a distinct time when evil began to change human existence from that "Eden-like" existence to one where selfishness seemed to rule?

THE FALL[70]

Clearly the fall, as we have just seen, is a central part of the biblical story. The question, however, is whether the Bible is *only* metaphorically describing the *ongoing* human condition, something that each person experiences as they lose the innocence of childhood. Or, alternatively, was there a literal fall, a real event or real set of events that initiated that downward spiral that led to human existence as we know it today? Some noted theologians and biblical scholars think that a faithful reading of Scripture does not lead one to conclude there was a specific event or set of events in literal history. For example, the noted Old Testament scholar Walter Brueggemann, in speaking of a distinctive fall, writes "Nothing could be more remote from the narrative itself."[71] Similarly for Karl Barth, there was "No Eden or Golden Age [which] stands at the head of human history."[72] For them, the biblical account is a profound metaphorical description of the human condition. However, Fretheim states that those who hold this view have been too quick to rule out the notion that Genesis 3 is also a "story of the past." Elaborating, he writes,

> Unless this is recognized, creation and fall are collapsed and God becomes essentially responsible for these adverse developments rather than the human being. It seems that "fall" language of some sort is truer to the text.[73]

69. Gen 4:7.

70. I am using the term "fall" in a narrower sense than some view the concept. For example, John Calvin gave it a cosmic dimension when he wrote: "The inclemency of the air, frost, thunders, unseasonable rains, drought, hail, and whatever is disorderly in the world, are the fruits of sin. Nor is there any other primary cause of diseases." (Cited in Provan, *Seriously Dangerous Religion,* 115.) In referring to this and other similar statements by Calvin, Iain Provan expresses strong disagreement. He writes, "This is precisely what the book of Genesis does not lead us to assume, nor does the remainder of the Old Testament." (Provan, *Seriously Dangerous Religion,* 116.) A cosmic fall that is the cause of all natural evil is not biblical, many Old Testament scholars argue, and I do not consider it further here. For further discussion see for example Provan, *Seriously Dangerous Religion;* Osborn, *Death Before the Fall;* and Fretheim, *Creation Untamed.*

71. Brueggemann, *Genesis,* 41.

72. Ramm, *Offense to Reason,* 139.

73. Fretheim, *God and the World in the Old Testament,* 108.

Similarly, the theologian Bernard Ramm is adamant regarding some type of fall into sin at the headwaters of humanity:

> Generic Adam is generic sinner. And this means at the beginning of human history is a Fall. That Fall accounts for the miserable history of the human species. To repeat: the miserable things that human beings do are not residual animal behavior patterns but the sins of fallen creatures.[74]

These latter two statements (Fretheim's and Ramm's) are the views that I find most convincing from both a theological and scientific perspective. For our highly cooperative nature to have developed as it did, goodness would have to have characterized our development as a species for a protracted (multigenerational) period of time. But that is not the end of the story. We know from many examples in recent history how readily one evil leader or a small group of like-minded individuals can readily sway the orientation of a host of others to carry out a set of dark and malevolent deeds: evil, personified. Given what we know about the spread of evil, it is not difficult to picture a fall that led to growing disharmony and a disruption of the spirit of goodness that had been pervasive in a specific community living in a harmonious fashion for multiple generations.

The pervasiveness of sin in God's very good world seems scandalous. Why would God create a world in which the presence of evil was possible? This poignant question is addressed by Provan:

> Genesis 3 opens by facing this reality directly: evil has, indeed, entered the world. It introduces us immediately and surprisingly to a creature of God who is apparently not under God's sovereign control nor under human dominion but who has apparently already "gone bad." This shadowy figure is a creature; he is one of "the wild animals the Lord God had made" . . . [He] does seem to represent dark, personal, but nonhuman forces that Genesis 1–2 has not led us to expect can exist in [a] cosmos that is "good." In such a world, according to Genesis 1:3–9, even the primeval watery chaos has been put to useful, constructive work. We must immediately assume, then (although we do not yet have any explicit indication of this), that in the world that God has made it is possible for creatures to depart from the good and turn toward evil. There is free will in the cosmos, and, as Genesis 3 opens, at least one personal, nonhuman creature has exercised it—badly. This important truth will be explicitly

74. Ramm, *Offense to Reason*, 82.

> confirmed later in Genesis 3, when the first human beings exercise their own free will.[75]

The biblical view, according to Provan, is that evil enters creation from within, which was made possible by the God-ordained freedom with which creation is endowed.

James Stump and Chad Meister elaborate further on the entry of evil into God's good world:

> Theologians throughout most of Christian history have almost universally agreed that God created the world for a good and loving purpose, which included a community of morally good human beings. It has also been almost universally agreed on that this divine plan was impeded in some significant way and that restoration was required. The commonly used term to denote this impediment is the fall, and a term often used to indicate its source is original sin.[76]

Most scholars agree that the genre of Genesis 1–11 indicates that we should understand that it contains at least some metaphorical language. But, as mentioned above, there is good reason to believe that the Genesis creation account is revealing real truths that are grounded in real history.[77] William Lane Craig writes that although the author intends for the narrative to refer to events that really occurred, "those persons and events have been clothed in the garb of the metaphorical and figurative language of myth."[78] The French theologian Henri Blocher emphasizes this as well:

> Figurative language drawn from myths may perfectly well suit an essentially historical revelation. Such a combination of imagery (of whatever provenance) and a message about definite events is familiar in Scripture; one need only think of Ezekiel's allegories, of apocalyptic visions, and of many of Jesus' parables. It involves no tension. It should cause no embarrassment . . . The real issue when we try to interpret Genesis 2–3 is not whether we have a historical account of the fall, but whether or not we may read it as the account of a historical fall. The problem is

75. Provan, *Seriously Dangerous Religion*, 117.

76. Stump and Meister, *Original Sin and the Fall*, 1.

77. In an extensive analysis, Craig uses the term "primaeval history" to explain the genre of Genesis 1:1–11. See Craig, *In Quest of the Historical Adam*.

78. Craig, *In Quest of the Historical Adam*, 286.

> not historiography as a genre narrowly defined . . . but correspondence with discrete realities in our ordinary space and sequential time.[79]

So the question, looking through a combined scientific and theological lens, is not whether there was a historical fall leading to the existence of original sin—it is highly reasonable to think there was—but what are the truths that emerge from the Scripture describing it. In so doing, it is important not to overinterpret or import preconceived notions into (the term for this is "eisegesis") the scriptural account. One of the most prescient examples of eisegesis that greatly impacted theology is Augustine's doctrine of how original sin is passed on from Adam through subsequent generations. Many scholars point out that Augustine's view was grounded in his use of an inaccurate translation of Greek into Latin of a particular key verse of Scripture—Romans 5:12.[80] That translation (the Latin Vulgate) mistranslated a Greek word that means "because" or "since" and did so in a way that caused Augustine to think Paul had written that it was in Adam "*in whom*" all sinned. In essence, he thought that the sin of all future humanity was actually found *in* Adam. That mistake, many scholars think, led to Augustine's doctrine that Adam's original sin led to something that was passed on through subsequent generations in a manner that is analogous in some way to what today we would say is the transmission of genes. Here is what Augustine wrote:

> Nothing remains but to conclude that in the first man all are understood to have sinned, because all were in him when he sinned; whereby sin is brought in with birth and not removed save by the new birth . . . It is manifest that in Adam all sin, so to speak, en masse. Hence by that sin we become a corrupt mass—"*massa perditionis*."[81]

ADAM, EVE, AND THE TRAVESTY OF SINFULNESS

So even though Augustine's doctrine that sin is passed on through heredity[82] is an inaccurate way of understanding Paul's teaching in Romans 5:12, we are nonetheless left with one overarching question: How ought we think

79. Blocher, *Original Sin*, 49.

80. David Fergusson writes that the "ontological involvement of each person in Adam's first sin ought to be regarded as an eisegesis of the text" (*Creation*, 40)." See also Green, "Wesleyan View," 68–70.

81. Cited in Parker, "Original Sin," 53.

82. I.e., sperm and/or eggs.

about the historicity of Adam and Eve? Was their existence metaphorical? Even though the genre, as Craig and many others state, leads to the conclusion that there is metaphorical language in the Genesis story, that alone does not address the question of whether Adam and Eve themselves were real individuals who lived in real history. We might for example reason that the entry of sin into that ancient community we have been discussing was a communal affair—that the community gradually slipped deeper and deeper into a sinful way of living as described in Genesis 1–11. Adam and Eve, according to this view, may have been a metaphorical representation of what happened to that community as it sank from one where goodness had been its hallmark and peaceful living its modus operandi to one where selfishness, sensual debauchery, and even hatred, ruled. Surely the downward spiral in the community could have started with just two individuals—Adam and Eve. But is this reading more into the story than would have been intended? Or alternatively, is it required to believe that there were just two persons at sin's headwaters? Is that what a faithful reading of Scripture requires? That is the question to which we now turn.

There have been at least four book-length analyses published in the last several years that lay out how scholars think about this question from both theological and scientific perspectives.[83] As I see it both the science and theology point towards a fall from a period during which goodness predominated to a time when sin entered the population. We know this as original sin and it continues to spread like a virus not through genetic transmission,[84] but as a cultural condition into which we are all born. But in addressing the question as to whether there were really two specific individuals responsible for the entry of sin, we turn to the writing of St. Paul. He is the one who, in the book of Romans, describes the entry of sin into the human condition especially poignantly.

As New Testament scholar Joel Green tells us, Paul provides three accounts of sin in Romans 1–7.[85] In the first (Romans 1:18–5:11), Paul tells "the story of the human family that chose creation over Creator." Paul's description of this in 1:18–32 is an "echo" of the first couple who "perverted [their] knowledge of God and sought to escape the status of creature." This establishes the pattern for idolatry, Paul writes, that would characterize Israel, and indeed all of humanity. The golden calf of Exodus 32, Green reminds us, would have been Exhibit A, but the downward spiral only goes on

83. McKnight and Venema, *Adam and the Genome*; Swamidass, *Genealogical Adam and Eve*; Haarsma, *When Did Sin Begin?*; and Craig, *In Quest of the Historical Adam.*

84. Although certainly genes that can lead to self-centeredness are present, as we discussed above in the context of David Sloan Wilson's work.

85. Green, "Wesleyan View," 152.

from there, "as they exchanged the truth of God for a lie, and worshipped and served created things rather than the Creator . . ."[86]

> Throughout this section of the letter, we find repeated references to sin as the outworking of *human* agency: *they* sin (2:12; 3:23), *they* take counsel with each other to do evil things (3:8), and *they* practice sinful behavior (1:32; 2:1–3).[87]

Having emphasized the scandalous role of human agency in allowing sin to corrupt the human experience, Green writes that the second account of sin is described in Romans 5:12–7:6. Here, Paul "identifies sin—or better Sin (a malevolent enslaving power)—as an agent influencing or causing humans to engage in sinful actions."[88] Green goes on to summarize Paul's position by writing that "Sin entered the world because of Adam, and Adam's disobedience had a ripple effect on the human family."[89] Thinking scientifically, one of the most important concepts that has emerged in current thinking about human origins is the role that culture has played in our evolutionary development as a species. In his pioneering book, the head of Harvard's department of human evolutionary biology, Joseph Henrich, writes, "the selection pressures created by culture can be among the most powerful observed in nature."[90] Our species has arisen through gene/culture co-evolution, Heinrich and many others argue. What they mean by this is that both factors—genes and culture—have worked in concert to shape human characteristics. Cultural habits can be passed both horizontally (through a population) and vertically (through generations). There are cultural activities (such as caring for the poor) that foster humanity's goodness, just as there are activities that cultivate sinfulness (such as gossip, habitual gambling, and pornography). Although Sin is not genetically inherited through sperm and eggs, it is a cultural force that spreads through time and space. It impacts the mind's orientation and thereby influences choices. Getting back to Green's analysis of the second portion of Romans (5:12–7:6), he writes that "Sin can be understood as an emergent entity, arising from human behavior while also holding human beings captive." Putting the two lines of thought together (Green's theological point and Henrich's anthropologically based conclusion), we can think of Sin (with that capital S) as a powerful cultural force

86. Rom 1:25.

87. Green, "Wesleyan View," 152 (emphasis mine).

88. Green, "Wesleyan View," 153.

89. Green, "Wesleyan View," 153.

90. Henrich, *Secret of Our Success*, 92.

that has been spreading through time and space ever since the proposed fall from goodness many millennia ago.

Although Sin is a force that spreads through culture, this does not imply that it emerges *solely* from within that culture as anthropologists largely assume. Scientific investigations, we should recall from chapter 1, are by the consensus of scientists themselves confined to the assumptions of methodological naturalism. This does not mean there is nothing outside the natural world influencing behavior, only that science does not have the "tools" to study it. Essayist Lance Morrow poignantly outlines how we ought to think about the source of what Green refers to as Sin. He writes:

> A crucial difference between wrong and evil is that people are implicitly in charge of the universe in which rights and wrongs are discussed; people have systems of laws to right wrongs. But evil implies a different universe, controlled by extra-human forces. Wrong is a human offense that suggests reparation is possible and deserved. Wrong is not mysterious. Evil suggests a mysterious force that may be in business for itself and may exploit human agency as part of a larger cosmic conflict—between good and evil, God and Satan.[91]

Similarly, in the memoir of the Canadian general who was in charge of the UN forces during the Rwandan genocide, we read:

> After one of my many presentations following my return from Rwanda, a Canadian Forces padre asked me how, after all I had seen and experienced, I could still believe in God. I answered that I know there is a God because in Rwanda I shook hands with the devil. I have seen him, I have smelled him and I have touched him. I know the devil exists, and therefore I know there is a God.[92, 93]

Having laid out Green's summary of Paul's first two accounts of human sinfulness in Romans, we now proceed to his analysis of the third (Rom 7:7–25). Paul begins this section by posing a question: "So what are we going to say? That the Law is sin?" To that, Paul replies, "Certainly not!" Here is Green's analysis of what follows in the next seventeen verses:

91. Morrow, *Evil*, 51.

92. Dallaire, *Shake Hands*, xviii.

93. For a masterful analysis of evil as it relates to the New Testament's portrayal of the significance of the crucifixion of Jesus, see Rutledge, *Crucifixion*, especially chapter 10.

> We find points of contact with Paul's account in Romans 1 as well as a key development in his perspective on Sin, grounded in what he has argued in Romans 5–6. Thus, the two actors in Romans 1 (humanity and God) are replaced by Sin and the law, with Paul's account turning on the ruse Sin plays on humans. Sharply put, Sin affects the way humans perceive the law (i.e., not by rejecting the law but by distorting human understanding of the law) and the way humans perceive their own behaviors . . .[94]

Green goes on to point out that Paul emphasizes that humans are not the victims here. They have chosen to sin and once they do that they come under the sway of Sin's power. (Paul made that clear in Romans 1.) Green goes on with his analysis of this third section of Romans:

> Paul's portrait of humanity concerns the entire human family—interconnected, interdependent, across time and space, with the sin of our earliest ancestors, we might say, spreading across all of humanity, like ripples radiating outward from a pebble tossed into a pond. Paul thus underscores the entire human family's solidarity in sin and interprets human sinfulness as a sign of human fragility and culpability, the result of which is that, without exception, all humanity stands in need of Christ's work.[95]

Paul wants his readers (and this includes us) to fathom that the ramifications of sin are devastating apart from the redemptive act of Jesus. Sin is a cancer that destroys the way that human interrelationships are meant to function. Paul concludes his distressing analysis of the human condition with this question: "Who will rescue me from this body of death?" The answer, "thanks be to God," is that it comes "through Jesus Christ our Lord."[96] Original sin is real, Paul tells us, and from this, humankind desperately needs to be rescued. So important is it to God's original intent in creating humankind that—as Paul writes in Philippians—God himself, through Jesus Christ, took upon himself "the very nature of a servant, being made in human likeness. And being found in appearance as a man, he humbled himself by becoming obedient to death—even death on a cross!"[97] In this Philippians passage Paul goes on to write, "Therefore my dear friends as you have always obeyed . . . continue to work out your salvation with fear and

94. Green, "Wesleyan View," 154.

95. Green, "Wesleyan View," 154.

96. Rom 7:25.

97. Phil 2:7b.

trembling, for it is God who works in you to will and to act in order to fulfill his good purpose."[98]

The sin problem, according to Paul (and indeed the entire body of Scripture), is the most significant challenge facing humankind. So serious is this travesty of God's purpose in creating humans to begin with, that God the Son became one of us to suffer an ignominious death, followed by his glorious resurrection to lead us into freedom from Sin's power. In Romans 5, Paul writes:

> For if, by the trespass of the one man, death reigned through that one man, how much more will those who receive God's abundant provision of grace and of the gift of righteousness reign in life through the one man, Jesus Christ! Consequently, just as one trespass resulted in condemnation for all people, so also one righteous act resulted in justification and life for all people. For just as through the disobedience of the one man the many were made sinners, so also through the obedience of the one man the many will be made righteous.[99]

Paul knows full well that sin entered not just through Adam but through Adam and Eve together. Indeed, for example, Eve's role is stressed in 1 Timothy 2:14. Paul depicts Adam as a real person who, through his actions, has brought sin into human existence, and he draws the parallel between the entry of sin by Adam to the solution of the sin problem through another person—Jesus Christ. So even though Adam is real in Paul's mind, he is not making the point that *only* Adam (and *not* Eve) is responsible for sin's entry into the human condition. Adam, in a sense, is a metaphorical representation of the first *couple*. It is both who are to blame. Paul knows this and even shows this by invoking Eve's culpability elsewhere. The point Paul is making is not that one person (as opposed to two) is responsible for sin's entry. Rather, the point is that Sin did enter, and that the life, death, and resurrection of Jesus was, and still is, the only solution to that catastrophic downfall. Given the extensive use of metaphor in the Genesis account, and Paul's own use of "metaphor" in using Adam as a representation for "the first couple," is anything to be lost from Paul's teaching if Adam and Eve themselves were—in Genesis—a metaphorical representation of a small community that is culpable for sin's entry? I think not. Speaking personally, I don't take a position on this matter, because (a) I don't think it matters

98. Phil 2:12, 13.

99. Rom 5:17–19.

theologically or scientifically[100] and (b) I don't know of any way of distinguishing between the two possibilities.

Paul also addresses this matter in 1 Corinthians 15, one of the Bible's most profound chapters because it centers so clearly on the foundational basis of Christian faith. He writes, "For as in Adam all die, so in Christ all will be made alive."[101] Again, Paul knew that according to the Genesis account Adam and Eve *both* set the wheel of human sinfulness in motion. He chose to cast Adam as the "metaphorical" representative of Sin's entry. The entry of sin into human existence has had devastating consequences and there is only one solution to the wreckage that Sin leaves behind. This is where our focus needs to be—on the single solution—Jesus—to humanity's single problem.

We were created to live into the presence of our Creator and to be ambassadors of God's love—as we care for each other and for creation as a whole. We must not allow ourselves to get bogged down in the details as to what is metaphorical and what is not in those first eleven chapters of the Bible. The main thing is that God created humankind, and as long as humans were living within the boundaries of their God-given role, it was not just good, but very good indeed. At some point in both time and space disobedience entered the human race and we continue to see the ramifications of sin's consequences each and every day. There is only one answer to this dilemma, but that answer is remarkable. The true human life, the way we are really intended to exist, was lived out before us two thousand years ago and the journey towards that existence for each one of us can begin today.

OMEGA: THE END

One of the most common questions that I receive after giving a seminar on human evolution is "Are humans still evolving?" The question is posed as though human evolution is driven by the process of natural selection, as though we are being shaped by natural influences over which we have little control. The fact is that for many thousands of years the nature of our existence has been molded and reshaped to an unprecedented degree by cultural forces working in concert with gene changes. The cultural phenomenon of language, for example, would have led to significant advantages to any genetic changes that led to alterations in the brain or vocal anatomy that enhanced language skills. Behavioral changes influence which genetic

100. See Swamidass, *Genealogical Adam and Eve,* for a detailed discussion of scientific considerations.

101. 1 Cor 15:22.

alterations will be selectively advantageous. Certainly, this gene/culture interaction is not unique to humans—many animals have some sort of culture[102]—but as humans came to fully understand the ramifications of others having a mind like their own, the significance of cultural influences on shaping the genome expanded exponentially.

Genes and cultural behavior change at vastly different rates. Think, for example, about the invention of the iPhone and its role in stimulating social phenomena such as Twitter, TikTok, Facetime, the QR code, GPS mapping, Uber, and YouTube videos. Invented less than two decades ago, it has changed not just how we communicate but literally all aspects of our lives. Who could have imagined in 2007 all that was on the horizon with the invention of this device? Today, in advanced economies like Japan and Argentina, a median of 75 percent of adults have a smartphone, but even in emerging economies like Tunisia and Indonesia the number approaches 50 percent.[103] The use of the smartphone and all of its ancillary functions is a cultural phenomenon that has changed human behavior in the blink of an eye compared to the rate at which a gene-change spreads through the population. Typically, this takes thousands of years.

There is another cultural change that has been emerging—albeit a little more "under the radar"—and that is our ability to manipulate genes. I began my career in genetics as an assistant professor in 1976, just a few years after the invention of recombinant DNA. This is a technique that enables investigators to combine genes from different organisms into single DNA molecules. It quickly became routine to amplify these hybrid molecules by a factor of billions and then trillions. In 1982, I led an interdisciplinary honors seminar at Syracuse University that focused on the potential of this technology for genetic engineering. At that time, bacteria were being used through this recombinant DNA technology to churn out a human protein, insulin, but there was much more on the horizon. The students and I marveled at projections that someday this would become a multibillion-dollar industry. Sure enough, by 1996, fourteen years later, its market capital had expanded to 7.6 billion dollars per year.[104] Then, by 2021, the global market value of the biotechnology industry had grown to 1,023 billion dollars a year, with a projected expansion to 2,550 billion by 2030.[105]

Unlike the smartphone, which is purely a cultural phenomenon, this development—also grounded in culture—is different. It provides the

102. See, for example, Ackerman, *Bird Way*, or Safina, *Becoming Wild*.

103. Pew Research Center, "Smartphone Ownership Is Growing Rapidly."

104. U.S. Dept. of Commerce, "Meeting the Challenge."

105 Grandview Research, "Biotechnology Market Size."

opportunity to guide our own evolution. Until now, given the slow pace by which new versions of genes spread through the population, our bodies have remained relatively unchanged over the past couple of hundred thousand years; they are still largely adapted to that hunter-gatherer lifestyle that characterized human existence for so long. But now we are on the cusp of being able to massively speed up the evolutionary process. At this writing, genetic engineering technology has been primarily used to alter somatic cells—those cells in our body which do not contribute genetic information to the next generation. In that capacity, it is being developed to treat cancer, various blood diseases, and even inherited genetic abnormalities like cystic fibrosis and sickle cell anemia. These studies have shown that this technology is going to provide relief for millions of individuals who suffer from the consequences of a vast array of diseases. However, with this capability comes the technology needed to also alter the genes of germ cells—the cells which give rise to sperm and eggs—and doing this would change the genetic makeup of subsequent generations. This would be taking human evolution into our own hands, and it raises significant ethical concerns: when, if ever, do we have the right to manipulate the genome of individuals not yet born? The fact is that once we start manipulating the germline and trying to direct human evolution there is a danger that we will unleash unforeseen effects.

In 2018, the first experiment that included changing the germline was carried out in China by a biophysicist, He Jiankui.[106] Three embryos containing an engineered gene thought to confer HIV resistance were implanted in women in the Chinese city of Shenzhen. There was a vast public outcry that anyone would attempt to enhance the human genetic makeup in this way. For one thing, the technique had not yet been sufficiently refined such that it could ensure that it would target only the HIV-sensitive gene. There was a danger that other—unanticipated—genes would be altered. In addition, it seemed doubtful that the parents, even though they had given their consent, were fully aware of the extent of the risk to which their future children (and grandchildren) were being subjected. As it turns out, the babies were born and now, several years later, little is publicly known about the health or even the identity of the toddlers. Arrested for not having gone through the proper regulatory channels, Dr. He, at this writing, has just completed a three-year prison term.

Here in the United States, one year prior to Dr. He's experiments, by the National Academies of Science and Medicine released a 265-page report of the results of a study of the ethics of human genetic engineering.[107]

106. Zimmer, "CRISPR, 10 Years On."

107. National Academies of Science and Medicine, "Human Genome Editing."

It did not rule out the prospects of genetic enhancement someday, but it urged extreme caution, including much public input before any modification of the human germline is attempted. Indeed, at the present time any such activity is currently illegal. But the basic technology has been invented, and it seems highly likely that it will be used—be it by a small rogue group working somewhere in the world, or by a government program designed to enhance the genetic capability of its people in order to gain economic and political power over its rivals.

Once we start taking matters into our own hands, unanticipated consequences can quickly multiply. Consider the following as a particularly poignant example. For a few decades a straightforward test has made it possible to determine if a specific fetus has Down syndrome. The test has enabled parents to make a decision as to whether they want to follow through with the pregnancy if the fetus has this genetic abnormality. In the US, until recently about 75 percent of Down syndrome pregnancies were terminated. In some countries, such as Iceland, the United Kingdom, and Denmark, the percentage of terminations is even higher—90 to 98 percent. So almost all pregnancies are terminated in those countries, but not quite all. Occasionally, some mothers opt to have their baby despite the diagnosis; in other cases, Down children are born because of an inaccurate test result. For example, in Iceland typically about two children are born with the syndrome per year. As a result, there are a small number of Down children in the population, and they will likely grow up to learn that almost all parents had chosen not to permit an individual like them to live. In a recent documentary, a thirty-year-old Icelandic woman who has Down syndrome was interviewed:

> Interviewer: For people who are listening and watching what would you tell them about people who have Down syndrome.
>
> Young woman: It's hard for me to say. [Long pause.] They just see Down; they don't see me.
>
> Interviewer: How does that feel?
>
> Young woman: It doesn't feel good. I want people to see that I am just like everyone else.[108]

This young woman had recently moved out of her childhood home and was now living with a partner. As the interviewer proceeded to have a conversation with the Down woman's mother in English, the young woman and her partner were present around the same table and across from each other. Not understanding the kitchen table conversation taking place in English between the reporter and the mother, the couple's focus turned

108. CBSN, "What Kind of Society?"

quietly to each other. As the interview continues, they can be seen forming hearts with their fingers, pointing lovingly back and forth across the table to each other. It is a sweet, loving sideline to the interview itself. So, here is the question: Is removing individuals like this from the human population, going to be a societal enhancement? I think not.

An Icelandic scientist, Kari Stefansson, was interviewed as part of the documentary. His comments are important partly because he has been a pioneer in genetics: his work has been especially significant worldwide in identifying genetic abnormalities that have specific health consequences. Here is an excerpt from that interview:

> Stephenson: My understanding is that we have basically eradicated, almost, Down syndrome from our society—that there is hardly ever a child with Down syndrome in Iceland anymore.
>
> Interviewer: What does the 100 percent termination rate, do you think, reflect about Icelandic society?
>
> Stephenson: It reflects a relatively heavy-handed genetic counseling. And I don't think that heavy-handed genetic counseling is desirable. . . . You're having impact on decisions that are not medical, in a way. I don't think there's anything wrong with aspiring to have healthy children, but how far we should go in seeking those goals is a fairly complicated decision.[109]

Once we start engineering the course of who gets to be born and what genetic qualities they will have, we may have a significant effect on the social dynamic in a way that *causes* hurt, rather than *heals* it. The Canadian bioethicist Francoise Baylis, referring to the genetic engineering of humans, has said, "I'm skeptical about the depth of understanding about what's at issue there. There's a difference between making people better and making better people."[110] We are treading on very hallowed ground as we rapidly move towards gaining the ability to direct our own evolution through techniques like using the gene-editing tool, CRISPR.[111] As mentioned above, in all likelihood there will be persons (and perhaps whole countries) who will seek to use these new technologies to manipulate the human genome, changing it in a manner that conforms to their idea of "better people." In other words, humans are about to embark upon a new era in which they will

109. CBSN, "What Kind of Society?"

110. Zimmer, "CRISPR, 10 Years On."

111. The above *New York Times* article by Carl Zimmer provides a fine overview of the CRISPR technology.

have the *knowledge* of how to direct their own evolution, but possibly not the *wisdom* regarding the many ramifications.

Interestingly, one of the most prescient books that deals with the ramifications of this topic was written over two hundred years ago: Mary Shelley's *Frankenstein.* She describes an artificially created human who begins his life as a sincere, friendly individual, but who, because of his physical appearance, doesn't fit in. Like all human beings, though, he does have a built-in need to be loved and accepted, and this is made even more acute because of his physical abnormalities and the resulting loneliness. When it becomes clear to him that he will never be accepted by society—that there is no place he can go to feel loved—life, for him, starts to spin out of control. It descends into utter chaos. We are entering an era where that story—the scientific fiction of our past—may soon become the nonfiction of our present. As we know from the experience of Dr. He, it is no longer unlikely that our *sicut Deus* inclinations will soon lead to someone else, or some group—seeking to "make a name for themselves"—to again attempt to manipulate human genes in a manner that they think will make a better human. Of course, countries can regulate such procedures through laws, but it is likely that not all countries will come to an agreement as to what ought to be permissible. But even if they do, the possibility that rogue individuals or groups will attempt to use this technology in their lust to manipulate will loom large.

So, we are on the verge of having the capacity to manage the evolutionary process, and with that will come the temptation to build changes into the genome to suit our own personal ideals. Be it greater intelligence, longer life, athletic prowess, musical genius, or physical beauty, parents will be tempted to order particular qualities for their children, and national leaders will be tempted to assemble a population of individuals who have qualities superior to those of their imagined enemies. We live in a fallen world, and if there is anything the past 150 years has shown us, no matter how good our intentions are at the societal level, sinful ambitions will always be crouching at the front door, waiting for a chance to take over the whole house.

It does not have to be that way, however. We now have a growing understanding of how our species came to be, and it is highly likely that it was our ancestors' escalating skills at cooperation—their ability to live lovingly—that has been the secret of our species' success. Biblical theology stresses that "God is love." First John 4:16 states, "Whoever lives in love lives in God and God in them." The origin of human beings is to be found in the God who is love. It was God's love for us and our love for each other that brought our unique form of life into existence. But sinfulness crept into our existence and distorted that original purpose. So determined was God that we be enabled to live as we were first intended to live that God became one of us—fully human while also fully God. God showed us and taught us what it means to live a life of love. More than that, however, God promised to empower us to live in a manner that is increasingly committed to our intended purpose. What does a life like that entail? One word permeates all that the life of Jesus depicts. That word is compassion. When the Gospel of John tells us of Jesus' response to the grief of Mary and Martha, it is too easy to overlook the significance of the two words—"Jesus wept." Despite his power to change the situation in that moment, he chose to enter into it, to feel and experience what the two sisters felt. He who is fully God shed tears of grief right along with those who were experiencing the excruciating agony that death of a loved one brings. What this example illustrates, then, is that to be fully human in the original sense of that term is to help carry one another's burdens. In the book of Matthew Jesus told a parable grounding it in life's bottom line. He made the puzzling assertion that what we do in caring for the hurts of one another, we also do for him.[112] It is far too easy to pass over these words without pausing to ponder their significance. What does Jesus mean when he said "Whatever you did for one of the least of these brothers and sisters of mine, you did for me"? Does this not mean that the Jesus who wept as he sensed the agony of Mary and Martha, in some sense continues to feel the agony of those who are suffering? If so, Jesus was saying that by caring for the suffering of others, we are, in some mysterious sense, bringing a degree of healing into the brokenness felt by the One who suffers alongside us.[113, 114]

112. Matt 25:34–46.

113. Consider Romans 8:26: "In the same way, the Spirit helps us in our weakness. We do not know what we ought to pray for, but the Spirit himself intercedes for us through wordless groans."

114. I am aware that it is much more nuanced than this. God, who exists outside of history, does not suffer as we finite beings suffer and furthermore God has no needs in a manner that parallels our needs, but it is helpful for me, as a finite being, to put it into these terms so that I can fully grasp its ramifications. Furthermore, I think this is why Jesus phrased the parable as he did. Here is a much more nuanced theological expression of the issues by the highly respected Catholic theologian, Herbert McCabe: "If the creator is the reason for everything that is, there can be no actual being which does not have the creator at its centre holding it in being. In our compassion we, in our feeble way, are seeking to be what God is all the time: united with and within the life of our friend. We can say in the psalm 'The Lord is compassion' but a sign that this is metaphorical language is that we can also say that the Lord has no need of compassion; he has something more wonderful, he has his creative act in which he is 'closer to the sufferer than she is to herself.'" (McCabe, *God Matters*, 44)

Jesus came to earth so that we might be transformed through the renewing of our minds into the original purpose for which we were created.[115] Sin's effect on humankind can have devastating consequences, but there is a cure. Just as we were formed by living lovingly in union with each other through the presence of God's Spirit, so Jesus has shown us that this is our ultimate destination. The journey is not without its challenges. The book of Acts makes this clear. We can expect pressures of all sorts to abound. So also, Paul's letters and the other epistles emphasize that we should not expect an idealistic existence. Not now. Not yet. But Paul writes that as new creations we can bear the first fruits, as compassion for the needs of others becomes our hallmark just as it was for Jesus. We are made to be loving co-operators; it was this characteristic that led to our origin as a species. The church, as Paul describes it, is meant to move us toward that type of existence.

So, let's turn back to the question that is frequently posed in seminars on human evolution. Are humans still evolving? As mentioned above, we have evolved to the point that we will soon have the skills needed to genetically control our evolutionary pathway. There are genetic diseases that cause human suffering more agonizingly painful than most of us can imagine. We will be able to use our knowledge of gene modification to end this suffering, bringing relief to those who are hurting—a very Jesus-like thing to do. As such, gene modification is a wonderful tool and there is every reason to embrace it as a means of caring for one another. However, this genetic knowledge also will facilitate attempts to *enhance* the human body, including the brain. This is where things can get messy as depicted so well in books like *Brave New World*, movies like *GATTACA*, or even in today's real-world experience as represented by the overzealous approach of some genetic counselors in Iceland.

In a recent conversation about her book *Altered Inheritance*, bioethicist Francoise Baylis talked about how we are going to be tempted to rush into modifying our genetic makeup. Recall that our species emerged and became well adapted for life in small cohesive groups in Africa. With that in mind, Baylis said, "I think you're going to hear scientists at some point telling us about the ways that we can modify ourselves to live in this environment, and at that point I am going to start screaming loud and clear 'Really? Why would we modify ourselves to live in this environment instead of changing our behaviors around our environment?'"[116]

115. See Rom 12:2.

116. https://www.youtube.com/watch?v=fGvocYptluo See 12:00 minutes to 12:17 minutes.

This behaviorally based approach is the entire thrust of David Sloan Wilson's book, *This View of Life*. In this book, he proposes that humanity needs to learn the lesson that evolutionary biology has taught us. We have been brought into existence by selection for those groups that cooperated most effectively, he writes, and we need to alter our behavior to function in that way once again. As an example of how we should function he cites a tract from an Anabaptist community of almost four hundred years ago:

> True love means growth for the whole organism, whose members are interdependent and serve each other. That is the outward form of the inner working of the Spirit, the organism of the Body governed by Christ. We see the same thing among the bees, who all work with equal zeal gathering honey.[117]

This is (essentially) straight out of 1 Corinthians 12, where Paul writes:

> Just as a body, though one, has many parts, but all its many parts form one body, so it is with Christ. For we were all baptized by one Spirit so as to form one body—whether Jews or Gentiles, slave or free—and we were all given the one Spirit to drink. Even so the body is not made up of one part but of many.[118]

Wilson goes on to emphasize that human evolution is not restricted to that which is brought about by genetic changes. For humans, we must consider the influence of social factors.

> Even though evolutionary biologists were gene-centric for most of the twentieth century, they have now gone back to basics by defining evolution in terms of variation, selection, and heredity, with genes as only one of several mechanisms of heredity. Cultural transmission represents another mechanism of heredity in many species, one that has become vastly more powerful in humans with the advent of symbolic thought . . . our species represents a new (or at least greatly expanded) evolutionary process.[119]

Wilson goes on to describe how the future of our species will need to consciously build upon what evolutionary science has taught us about how our species came to be. What we must do now, he writes, is to consciously broaden our focus on the welfare of the small group to one that encompasses

117. Ehrenpreis et al., "An Epistle on Brotherly Community," cited in Wilson, *This View of Life*, 90.

118. 1 Cor 12:12–14.

119. Wilson, *This View of Life*, 219.

the welfare of the whole planet. That, he states, needs to be the chief moral criterion—the compass by which we define good and evil. Note that he is not talking about changing genes; he is talking about changing behavior.

As mentioned before, Wilson considers himself an atheist, but what he is really doing is taking the biblical description of the body of Christ and emphasizing that this is how we came to be a successful biological species to begin with. Given that we now live in an interconnected global community, this model, he states, needs to be expanded. In essence, we need to enlarge the community for which we care so that it includes all of creation.[120] The whole earth is God's very good creation. What made it "very good," and not just "good," is that humans were here, living as God intended for them to live as God's representatives. In order to be better people, we do not need genetic enhancement to evolve into those better people. What we need is to be new creations. We need to be attuned—spiritually—to all that God is calling us to be as we carry out our purpose to live as God's representatives. It is to this end that we have been created, and it is to this end that we can be empowered to live.

120 Wilson is very explicit that his ideas are largely inspired by the Christ-centered writing of the Catholic priest Teilhard de Chardin.

Afterword

By Cara Wall-Scheffler and Robert Wall

CMWS:
Professor Falk very kindly asked me (Cara Wall-Scheffler, PhD in biological anthropology) and my father (Rob Wall, ThD in biblical studies) to respond to his book in the form of this Afterword. In our minds, the job of an afterword is to use the ideas and data from the volume to springboard into dialogue and discussion covering what we learned, and where we can go from here. While we refer to many of Professor Falk's arguments and ideas with specific examples below, we also hope that you have taken his ideas to heart and will see our expansions on his points even when we do not directly cite them. Additionally, we have been asked to engage in dialogue as if it were in fact a dinner conversation between us—a request with which we were pleased to comply as dinner conversations are an important aspect of our family's culture. We trust that your own family will engage in this work together, and that the book will make the rounds as your own loved ones grapple with the complexity of God's created world, our particular place in it, and how we can understand God's love throughout.

RWW:
Professor Darrel Falk's monograph is an exemplary example of what a conversation between science and theology looks like at ground level. The conclusions he draws from his scientific research in genetics, freighted as they are by the science of human evolution, are presented for us to review as good science. Likewise, his constructive reworking of Christian theology's pivot point—that humans can enjoy an existential relationship with the transcendent triune God—is based upon careful reflection of Christian biblical interpretation and, again, presented to us to review as good theology. An honest dialogue between the two, theology and science, begins

here with good science and good theology engaged in an interpenetrating dialogue. If we agree with St. Augustine that all truth is God's truth, then the ends of this dialogue in the academy is a thicker, fuller understanding of human evolution made more proximate. This axiom does not require us to agree completely with each other about our theological understanding or scientific conclusions, nor about how the two intersect in mutually informing ways. We should, however, agree that theology and science are worthy partners in a conversation about truth.

The two most consequential canons or universal plumblines in guiding the church's theological understanding and so hidden partners in this dialogue between theology and science are the church's Scripture and its ecumenical creeds. Providentially, they were formed by the early church at roughly the same time, based upon the apostles' trustworthy memories of the historical Jesus whom they eyewitnessed and believed to be the very personification of truth as God's incarnate son. Both Scripture and creed make clear the cooperation between Father and Son in the creation of all things. The Nicene Creed (325 CE) not only begins by affirming God, the Father almighty, as "maker of heaven and earth, of all things visible and invisible" but goes on to affirm that through God's son "all things were made." Both affirmations are earthed in Scripture's narratives of creation, which are then interpreted and repeated in subsequent biblical texts. For example, the opening *exordium* of the New Testament book Hebrews plainly affirms that God's eternal son is the one "through whom the worlds are created . . . and by whom all things are upheld by his powerful word" (Heb 1:2–3). The prologue to the Fourth Gospel even more plainly claims that God's preexistent Word created all things and apart from this Word not a thing created was created (John 1:3). It is God's creating Word that "became flesh and took up residence among us" on planet earth as Jesus Christ (John 1:14–17).

Perhaps it is because of their role in creation, clearly affirmed in both Scripture and creed, that the Father and Son form a traditional orthodoxy as co-creators of all things; this partnership is the centerpiece of most Christian theologies of creation. True to the Wesleyan spirit that has helped shaped Professor Falk's theological instincts, his work has smuggled in God's Spirit into the dialogue between a fully trinitarian theology of creation and the science of human evolution. But in fact the Nicene Creed affirms that the Spirit is the "giver of life," an idiom that trades on the programmatic claim of John's Gospel that the Spirit "is making life" (*estin to zōopoioun*, John 6:63). The force of John's present participle, especially in contrast to the following emphatic denial that "making life" is the work of "the flesh" or human nature, may envision the Spirit's present or continuing agency in making human life—something that nature itself cannot do alone. While this reading of the Gospel saying offers potential in Scripture's support of Falk's thesis that the Spirit is involved in human evolution, it is hardly certain given its eucharistic setting in the Fourth Gospel and the characteristically cryptic tenor of Jesus' own commentary on the saying. Even the context of its creedal formulation, the life the Spirit makes seems related to its role in creating the church's existence as "one holy, catholic, and apostolic," as Falk notes in his exposition on Israel's formation.

Scripture's opening narrative of creation is a severely gapped story. Nonetheless, its essential plotline may be read as programmatic of God's way of salvation: In the beginning, earth was in chaos, shrouded in darkness and unable to bring forth life (Gen 1:1–2). Six days later, the Creator surveyed all that had been brought forth and "saw that it was very good" (1:31). What the reader recognizes in this mysterious story of how God the Creator transformed bad news into good news lies at the heart of Scripture's self-presentation of God the Savior. In between the beginning and ending of this story of salvation are those narrative gaps with slippery slopes that both theologians and scientists have attempted to fill in adding commentary to this canonical version of life's origins.

My brief comments in this Afterword will not rehearse the vast and varied history of this narrative's interpretation in both the synagogue and the church. I will only allow that a Christian *theological* reading of it must be guided by the canonical creeds; those mechanistic and relational gaps filled by modern scientists, including contemporary evolutionists such as Falk and Wall-Scheffler, must follow their own canons of scientific method. Without doing so, the prospective dialogue between theology and science cued by Scripture's programmatic story of origins will fall victim to unhelpful speculation and worse.

In this Afterword I wish to stay on the creedal commitment to the Holy Spirit as the "giver of life," since one of the contributions of Professor Falk's monograph is to draw attention to the neglected role the third person of the holy Trinity performs in the evolution of human existence. That is, he does not use evolution as a typology that helps explain church doctrine of the Holy Spirit; rather, he draws upon pneumatology (the study of the Holy Spirit) to help explain a critical moment in human evolution. A passing note in this regard. In the New Testament book, 1 John, the Elder who wrote this letter makes the stunning explanation of an even more stunning (and difficult) claim that God's children are unable to sin because they are "fathered" (*gennaō*) by God and thereby reborn with "God's DNA" (*sperma autou*, 1 John 3:9)—what Paul might call a "new creation" (2 Cor 5:17). Here Scripture reimagines salvation as the genetic or natural transformation of human nature! You have already read this thesis in Falk's manuscript, and we discuss the transformation further.

These ideas—and many more images of creation found in Scripture—are founded on the Bible's opening story. What its readers rarely pay attention to is this story's *inclusio*—those literary brackets that frame and guide its interpretation and implication. In the storyteller's opening description of a lifeless and formless earth on day one of creation, mention is made of God's hovering *pneuma* ("spirit"). The Hebrew word translated *pneuma* is *ruach*, a polyvalent word that may mean "spirit" or "wind" or "breath" depending on its context. Here it refers to God's Spirit. Critically, the verb attached to Spirit is *epipherō*, which according to Robert Alter's wonderful translation of the Hebrew text may envisage a bird's "fluttering"—moving air with its wings—over its nest and so carries the connotation of giving birth or nurture of its young.[1] The contrast between what a lifeless earth was at its beginning and yet the presence of a "fluttering" Spirit who gives and nurtures life could not be more stark but more evocative of a divine Spirit, "the giver of life." I would argue that every narrative "gap" of creation's severely gapped story that follows from Scripture's prefatory sentence is at least in part occupied by this fluttering, life-giving Spirit. Every mechanism of nature's process of selection in human evolution in some way works because of its presence.

In addition to the theological discussion Falk has presented here, Daniel Castelo offers a provocative exposition on the relationship between the Spirit and science that is worth considering as an extended footnote to this idea. Castelo challenges the idea of the Spirit "intervening" in nature to fill in the gaps or "blind spots" left momentarily unexplained by science by suspending natural processes—a dyadic conception that he finds unhelpful because it often keeps theology at arm's length from science. In fact, sometimes Christians appeal to divine revelation to explain what scientists of nature cannot yet explain; this amounts to a kind of nonscientific reductionism or apologetics that stops the dialogue between theology and science rather than encourage or guide it. I find this move, motivated often by a deep faith in God, encourages secular scientism rather than to open up the study of science to new insight about reality. Castelo himself prefers a kind of non-dualistic "pneumatological cosmology" that advocates for an "en-Spirited" or "enchanted" naturalism, trading on language used by the philosopher James K. A. Smith. This approaches the critical study of the real world or of creatureliness as the embodiment of both material and spiritual goods—an integral whole that must be studied as such by both theologians and scientists. This sort of slant toward biology and all its mechanisms of

1. Alter, *Five Books of Moses*, 11.

life are studied as "Spirit-drenched." The entire process of evolution, including mutation and adaptation, variation and inheritance, fitness, and so on, is Spirit-directed from origins into eternity (and perhaps continues there as well!). There is nothing of nature, including its various trade-offs, that is not graced and enabled by the Giver of Life.[2]

At the center of Professor Falk's idea is this: that the variety of successes in human evolution is to a certain extent determined by God's cooperating grace prompted or "made" through the Spirit's inward nudges. In this way, the Creator's Spirit, the "giver of life," may be thought of a divine mechanism of human life. In developing this hypothesis, Falk slips into the creation narrative to offer a biologist's explanation of the story's famously consequential line that tells readers of the triune God's "making a human (*'adam*) in our image as our likeness so they may hold sway (*radah*)" over all the other creatures God made (Gen 1:26). The question both theologian and biologist might ask of the story's climactic line is this: how is it that God made humans into creatures so successful that they would "hold sway" over all other creatures? Is there a genetic basis for thinking more carefully about what is needed for humans to "hold sway" over all other creatures and fill and till the earth with its progeny and produce?

Regarding the Spirit's role in this evolution, Falk slants the discussion somewhat differently than Paul does. Paul envisions the Spirit's primary role in reforming faithful believers into the likeness of God, which is the pivot point in his revolutionary rethinking of the role of God's Spirit in the economy of salvation. Consider, for example, the creational allusions of Paul's programmatic affirmation of Colossians 3:10 that the believer is "made new in knowledge (*epignōsis*) according to the image of (the God) who created humans." In his final synthesis of this idea of the Spirit, Paul writes that God transforms human existence into God's likeness "through the washing of this new birth and the renewing of the holy Spirit whom God poured out into us in full measure through Jesus Christ our savior" (Titus 3:5–6).

I take it that Falk's essential argument complements this Pauline idea. That is, God pours out God's Spirit into human lives so that they are made capable of responding to threats and opportunities with increasing insight and so are "fit" to evolve and survive if not thrive. Paul's idea of the kind of "knowledge" (*epignōsis*) mentioned in Colossians 3:10 that is communicated by God's indwelling Spirit seems very much like Falk's idea of humankind's ever-increasing awareness or intuition of what is truly true or an evolving knowledge of the implications of their choices. One might even imagine that humans come with a genetic makeup that would enable them

2. Castelo, *Pneumatology*, 65–80.

to receive and respond to the Spirit's communication—or "knowledge"—of God's intentions for being human. Perhaps it is even possible for us to reimagine Paul's catalogue of social virtues of the Sprit-transformed community "clothed with hearts of mercy, kindness, humility, gentleness, patience and forgiveness" (Col 3:12-13) as those same dispositions that not only draw us naturally into life with God but also with one another in forming a social solidarity "fit" for a particular kind of everlasting survival. Naturally, Scripture and especially Paul's letters are an "occasional" literature, written and still read as God's word for a particular group at a particular moment in time. The study of human evolution and its various mechanisms of life concern changes that take root from generation to generation. Nonetheless, Scripture discloses a way of thinking about the Creator's Spirit—really about a triune Creator God—that has potential in helping students reimagine the full sweep of hominin evolution into God's image according to God's likeness.

In concluding creation's story, the storyteller uses a second Hebrew word for "breath," *nephesh*, which is translated by the Greek word *psuchē*. This is another polyvalent word that may mean different things depending on context, but it generally refers to some characteristic that distinguishes a living or sentient creature. The phrase "breath of life" (Gen 1:30) marks out every living creature, whether human or nonhuman, that occupies a "very good" earth at the end of the sixth day of creation. In that both "breath" and "life" are features of the Spirit, this phrase may be thought of as a title of the Spirit's providential role in animating nature. What seems critical to me is that this Spirit-*inclusio* is open-ended. The grade given the Creator's salvation of a chaotic, lifeless earth is "very good" rather than perfect. One of the principal misconceptions of biological adaptation—that human evolution is "moving air" toward a finite telos—is based upon our bent to meet certain "learning goals" in the classroom or achieving goals in the workplace or completing "bucket lists" at the end of life. This way of daily life forges a pattern of thinking that requires an endgame toward which life is necessarily moving. We are motivated to meet present needs in order to advance in life. For this reason, most of us are bothered by the "very good" at the story's end. We much prefer the outcome of God's way of creation to be perfect.

But here is why this very good ending is, after all, a perfect one. Many scholars comment that the choice of "very good" instead of "perfect" anticipates the story of Eve and Adam's disobedience of God's command, which Christians (especially in the West) refer to as "the original sin" or "the fall." God's plan of salvation takes its cue from this moment of human failure. However, a closer reading of the opening narratives of Scripture—narratives that set the table for everything else that follows in the church's two-testament Bible—suggests that the judgment of a "very good" creation does not anticipate the story of failure that follows but it concludes a story that already envisions God's way of salvation. Indeed, God created the heaven and earth, every invisible and visible creature—all of which formed a very good interdependent and living whole—in hallowed response, in Alter's memorable translation, to an "earth (which) then was welter and waste and darkness over the deep"—that is, a futile place without life or light. Creation itself was a redemptive act of the Creator; God did not wait for the sin and failure of humans to save the world! And at every moment and in every gap the reader should assume the presence and partnership of the fluttering Spirit of God giving birth and nurturing life.

The kind of dialogue I have in mind would include faithful scientists of biological evolution who in conversation with theologians may imagine the nature of Spirit's work in the various mechanisms of selection. A "very good" creation is open-ended without a finite telos. Natural selection is on the move, itself pneumatic and open-ended, perhaps fluttering in the currents of the Wind or Spirit of God. A creation that is not "perfect" may allow for different levels of fitness and so the survival of some within a population of creatures but not of others unable to adapt to a changing and challenging environment. Certainly, the narrative of day six presumes the necessity of reproduction. And so on. My point is a theological one and not yet informed by the science of Professor Falk's exposition: nature's selection, the mechanism of life, may be thought of as a pneumatological process in which the "Giver of Life" is constantly at work in bringing order to disorder, trading life for death, giving breath to plants and animals but also taking it away in non-random process of holy survival.

Finally, the research question Falk puts forth here is the same one I also ask of this coalition of scientists and theologians: where and how might we see God's Spirit at work in any of these mechanisms of evolution (here Falk focuses specifically on humans)? This is a different sort of question than we sometimes invite when theologians use the science of evolution as typological of the economy of God. In this case, natural selection alone may be useful as an explanatory model of sorts of theological ideas that involve a formative or transformative process. This research question seeks a way of *integrating* a constructive pneumatology with the science of human evolution. How might the theologian's Spirit help the scientist explain certain features of biological evolution and how might the scientist's idea of biological evolution deepen the theologian's understanding of the Spirit as the Giver of Life?

CMWS:

In response to this question, I would note that during family dinner conversations, my father often tosses out numerous "teasers," each of which could themselves take a volume to appropriately play out. This said, there are two teasers that I will take on as part of our response to Darrel's wonderful book. First, as an evolutionary biologist I would elaborate the idea of biological perfection to observe that such "perfection" can simply never be achieved, even in our own bodies. Even a state of wellness (i.e., non-sickness) is a dynamic push and pull between all our different physiological systems. Maintaining homeostasis in one system often requires another system to be partly out of homeostasis, for example maintaining low blood pressure to protect the elasticity of your heart and aorta might result in low tissue oxygenation. Or, mechanisms that maintain blood levels of calcium are not interested in bone density, and may (permanently) borrow calcium from the bones in order for the heart to beat effectively. To this end, our own bodies expect a dynamic relationship where one system is briefly out of homeostasis to aid another system doing some other task. We can very quickly expand this situation up to the level of a niche or ecosystem, in which the interplay of organisms similarly requires a give and take for biodiversity to flourish. It is simply not possible in this particular creation to have both perfection and freedom. The goodness means that creation persists within the push and pull of the myriad of organisms interacting with each other and the abiotic features of any environment. Everything you could possibly imagine about our abiotic environment has changed: the amount of photons hitting the earth's surface bringing in energy, the congregation of land mass at the equator (or at the poles), the location of the magnetic north (or south), the time it takes the earth to circumnavigate the sun. Everything you could possibly imagine about the biotic environment has similarly changed: the nature of metabolism and energy creation, the organization of the cell membrane, the manner of reproduction, meiosis and mitosis. If the created order did not have the ability to adjust to the position of the moon, then there would be no life: there must be dynamic interplay between the units of creation for life to be ordered.

Second, I would also ask as a Christian theist how might God's Spirit "flutter" over Creation in real time? Professor Falk has dug deeply into mechanisms by which allele frequency has changed over our lineage, but I would like to stay located on the Wesleyan track of "warming" our hearts in our lifetime—that is, changing the very nature of our embodiment in a way that might still be heritable. This might allow us to consider the question posed above: how might salvation transform human nature and, subsequently, how might this pneumatic fluttering play out evolutionarily and maybe even via natural selection? Work in answering these questions both theologically and biologically may allow us to grow in our understanding of the wonders of creation that Professor Falk has so elegantly primed for us in his narrative up to this point. Genes 1 and 2, which Falk discusses in chapter 5, are, as he notes, extremely special—they change the regulation of many other genes, which is why when they change, there are many coordinated changes around the body. There is a growing body of evidence that how we behave changes the way in which our genes are regulated; Falk already reminded you that behavior often comes before allele frequency changes (chapter 3, Baldwin Effect) and here I am alluding to additional studies that focus on the field of epigenetics. A reading of this literature elucidates not only on how behavior shifts how genes are regulated in your life, but your behavior and environment can shift how your child's genes are also regulated.

Studies in this area often collect data in problematic areas: e.g., how certain harms might impact a body and that body's offspring. These harms can be both abiotic (extreme temperatures) and biotic (nutrition, disease), as well as social (poverty, prejudice). For example, women who experience discrimination due to racism are more likely to experience miscarriages and more likely to give birth to babies underweight (small birth size is then correlated with long-term health issues). These tragedies are not at all caused by their genes, but caused by how their genes are regulated, and thus what proteins cross the placenta during pregnancy. In other mammals it has been shown that malnutrition and other forms of stress can lead to permanent changes in gene expression that essentially act like mutations but are not—thus you are able to inherit changes to how your DNA is used, even without having to change the DNA itself.

A pneumatology of biology might animate such DNA expression. At this point in history, we are hoping for New Creation in which our capacity to make behavioral changes in coordination with the Spirit's guidance shift even further towards the Creator's full expectation for our lives—a "healing" of creation.[3] By accepting the Spirit's guidance, we change our behavior, thus changing our gene regulation, thus changing the biological outcomes for ourselves and our offspring. Given the number of studies investigating changes in neural tissue quantity and activity during prayer, investigating prayer and other health indicators is not outside the realm of reasonable hypothesis testing. Is it possible that the Spirit's "fluttering" over us enlivens our biology so that the alleles we do have our used in a healthy way? Does our Spirit-directed prayers to the Lord, the maker and sustainer of

3. Snyder, "Salvation Means Creation Healed."

all things, help offset the harm of discrimination or poverty, for example? Without delving into organized religion's "opiate" status, considering the actual biological changes of an active prayer life could help the scientist understand human evolution today.

As Falk notes, citing the work of Terence Fretheim and J. Richard Middleton (chapter 7), accepting the Spirit's guidance is the only way to bring forth New Creation. The earth itself is not to blame when earthquakes kill people. Our biology itself is not to blame when people die from viral infections. God is not to blame.[4] Instead, as Fretheim so clearly parses out in *Creation Untamed*, it is our failure to say yes to the Spirit that prevents New Creation and that keeps our bodies from reaching our "goodness." We must allow the Spirit to flutter over us so that we can again dynamically integrate with the rest of God's creation.

Cara Wall-Scheffler, PhD
Professor and Chair of Biology
Seattle Pacific University

Robert Wall, ThD
The Paul T. Walls Emeritus Professor of Scripture & Wesleyan Studies
Seattle Pacific University and Seminary

4. Castelo, *Theological Theodicy*.

Bibliography

Abraham, William J. *Crossing the Threshold of Divine Revelation*. Grand Rapids: Eerdmans, 2006.

Ackerman, Jennifer. *The Bird Way: A New Look at How Birds, Talk, Work, Play, Parent and Think*. London: Penguin, 2020.

Alter, Robert. *The Five Books of Moses: A Translation with Commentary*. New York: W. W. Norton, 2004.

Augustine. *The Confessions*. Translated by Maria Boulding. Hyde Park, NY: New City, 1997.

Aujoulat, N., et al. "La grotte ornée de Cussac (Dordogne), observations liminaires." *Paléo* 13 (2001) 9–18.

Barrett, Justin L. *Born Believers: The Science of Children's Religious Beliefs*. New York: Free, 2012.

Bauckham, Richard. *Jesus and the Eyewitnesses: The Gospels as Eyewitness Testimony*. 2nd ed. Grand Rapids: Eerdmans, 2017.

BBC Radio. "The Beginnings of Belief." *Living with the Gods*, October 24, 2017. https://www.bbc.co.uk/programmes/b099xhmj.

Becker, Ernest. *The Denial of Death*. New York: Free, 1973.

Behne, T., et al. "Twelve Month Olds' Comprehension of Pointing." *British Journal of Developmental Psychology* 30 (2012) 359–75.

Berger, Lee. *Almost Human: The Astonishing Tale of* Homo naledi *and the Discovery that Changed Our Human Story*. Washington, DC: National Geographic, 2017.

Berger, Lee R., et al. " *Homo naledi* and Pleistocene Hominin Evolution in Subequatorial Africa." *eLife* 6 (2017) e24234.

Berkowitz, Leonard. *Aggression: Its Causes, Consequences, and Control*. Philadelphia: Temple University Press, 1993.

Berry, R. J. "Did Darwin Dethrone Humankind?" In *Darwin, Creation and the Fall*, edited by R. J. Berry and T. A. Noble, 30–74. London: Inter-Varsity, 2009.

Bliss, Catherine. *Social by Nature: The Promise and Peril of Sociogenomics*. Stanford, CA: Stanford University Press, 2018.

Blocher, Henri. *Original Sin: Illuminating the Riddle*. Downers Grove, IL: InterVarsity, 2000.

Bloom, Paul. "Religious Belief as an Evolutionary Accident." In *The Believing Primate: Scientific, Philosophical, and Theological Reflections on the Origin of Religion*, edited by Jeffrey Schloss and Michael Murray, 118–27. Oxford: Oxford University Press, 2009.

Bonhoeffer, Dietrich. *Creation and Fall: A Theological Exposition of Genesis 1–3*. Ebook ed. Dietrich Bonhoeffer Works 3. Minneapolis: Fortress, 1997.

Browne, Janet. *Charles Darwin: The Power of Place*. Princeton: Princeton University Press, 2002.

Brueggemann, Walter. *Genesis*. Interpretation: A Bible Commentary for Teaching and Preaching. Louisville: Westminster John Knox, 1986.

Brusatte, Stephen L. "What Killed the Dinosaurs?" *Scientific American* 313.6 (2015) 54–59.

BvonHoldt, Bridgett M., et al. "Structural Variants in Genes Associated With Human Williams-Beuren Syndrome Underlie stereotypical Hypersociability in Domestic Dogs." *Science Advances* 3 (2017) e1700398.

Callaway, Ewen. "Oldest *Homo sapiens* fossil claim." https://www.nature.com/news/oldest-homo-sapiens-fossil-claim-rewrites-our-species-history-1.22114.

Capra, F., and P. L. Luisi. *The Systems View of Life: A Unifying Vision*. Cambridge: Cambridge University Press, 2014.

Carroll, Sean B. *A Series of Fortunate Events: Chance and the Making of the Planet, Life, and You*. Princeton: Princeton University Press, 2020.

Castelo, Daniel. *Pneumatology: A Guide for the Perplexed*. London: Bloomsbury, 2015.

———. *Theological Theodicy*. Eugene, OR: Cascade, 2012.

Castelo, Daniel, and Robert W. Wall. *The Marks of Scripture: Rethinking the Nature of the Bible*. Grand Rapids: Baker Academic, 2017.

CBC Radio. "Interview with Steven Shapin." *How to Think about Science*. http://www.cbc.ca/radio/ideas/how-to-think-aboutscience-part-16-1.464997.

CBSN. "What Kind of Society Do You Want to Live In? Inside the Country Where Down Syndrome Is Disappearing." *On Assignment*, August 15, 2017. https://www.cbsnews.com/news/down-syndrome-iceland/.

Charrier, Cécile, et al. "Inhibition of SRGAP2 Function by its Human-specific Paralogs Induces Neoteny During Spine Maturation." *Cell* 149 (2012) 923–35.

Conway Morris, Simon. *Life's Solution: Inevitable Humans in a Lonely Universe*. Cambridge: Cambridge University Press, 2003.

Cooperrider, Kensy, et al. "The Protean Pointing Gesture: Variation in a Building Block of Human Communication." *Proceedings of the Annual Meeting of the Cognitive Science Society* 36 (2014) 355–60.

Cottingham, John. *Why Believe?* Ebook ed. London: Continuum, 2011.

Coyne, Jerry A. *Faith versus Fact: Why Science and Religion Are Incompatible*. New York: Viking, 2015.

———. "Reflections on Life of Richard Lewontin." July 5, 2021. https://whyevolutionistrue.com/2021/07/05/dick-lewontin-1929-2021/.

Craig, William Lane. *In Quest of the Historical Adam: A Biblical and Scientific Exploration*. Grand Rapids: Eerdmans, 2021.

Crick, Francis. *What Mad Pursuit: A Personal View of Scientific Discovery*. New York: Basic, 1988.

Dallaire, Roméo. *Shake Hands with the Devil*. Boston: Da Capo, 2004.

Danchin, E., et al. "Cultural Flies: Conformist Social Learning in Fruitflies Predicts Long-Lasting Mate-Choice Traditions." *Science* 362 (2018) 1025–30.

Darwin, Charles. *The Autobiography of Charles Darwin*. Edited by Nora Barlow. London: Collins, 1958.

———. *The Descent of Man, and Selection in Relation to Sex*. Kindle ed. London: Penguin Classics, 2014.

———. *On the Origin of Species by Means of Natural Selection, or Preservation of Favoured Races in the Struggle for Life*. London: John Murray, 1859.

Dawkins, Richard. *The Ancestor's Tale: A Pilgrimage to the Dawn of Life*. Boston: Houghton Mifflin, 2003.

———. *The God Delusion*. Boston: Mariner, 2006.

———. *The Selfish Gene*. Oxford: Oxford University Press, 1989.

De Chardin, Teilhard. *Hymn of the Universe*. New York: HarperCollins College, 1969.

———. *The Phenomenon of Man*. Torchbook ed. New York: Harper, 1976.

De Queiroz, Alan. *The Monkey's Voyage: How Improbable Journeys Shaped the History of Life*. New York: Basic, 2014.

De Waal, Franz. *Are We Smart Enough to Know How Smart Animals Are?* New York: W. W. Norton, 2017.

———. *Mama's Last Hug: Animal Emotions and What They Tell Us About Ourselves*. New York: W. W. Norton, 2019.

Deacon, Terrence. "On Human (Symbolic) Nature: How the Word Became Flesh." In *Embodiment in Evolution and Culture*, edited by Christian Tewes and Gregor Etzelmüller, 129–45. Tubingen: Mohr, 2016.

———. *The Symbolic Species: The Co-evolution of Language and the Brain*. New York: W. W. Norton, 1998.

Deacon, Terrence, and Tyron Cashman. "The Role of Symbolic Capacity in the Origins of Religion." *Journal for the Study of Religion Nature and Culture* 3 (2010) 490–517.

Dennis, Megan Y., et al. "Evolution of Human-specific Neural SRGAP2 Genes by Incomplete Segmental Duplication." *Cell* 149 (2012) 912–22.

Desmond, Adrian, and James S. Moore. *Darwin: The Life of a Tormented Evolutionist*. New York: W. W. Norton, 1991.

———. *The Descent of Man. Selection in Relation to Sex by Charles Darwin, 1871. (Introduction)*. London: Penguin Classics, 2017.

Dobzhansky, Theodosius. *Mankind Evolving*. New Haven: Yale University Press, 1962.

Dos Reis, Mario, et al. "Phylogenomic datasets provide both precision and accuracy in estimating the timescale of placental mammal phylogeny." *Proceedings of the Royal Society B* 279 (2012) 3491–3500.

Dunbar, Robin. *Human Evolution*. Oxford: Oxford University Press, 2016.

Ebersberger, Ingo, et al. "Genome-Wide Comparisons of DNA Sequences Between Humans and Chimpanzees." *American Journal of Human Genetics* 70 (2004) 1490–97.

Ecklund, Elaine Howard. *Science vs. Religion: What Scientists Really Think*. Oxford: Oxford University Press, 2010.

Ehrenpreis, A. C., et al. "An Epistle on Brotherly Community as the Highest Command of Love." In *Brotherly Community: The Highest Command of Love*, edited by R. Friedman, 978. Rifton, NY: Plough, 1978.

Enns, Peter. "What Does Image of God Mean? Part I. Not the Soul." BioLogos, 2010. *https://biologos.org/articles/what-does-image-of-god-mean/*.

Ewen, C. "Oldest *Homo sapiens* Fossil Claim Rewrites Our Species' History." *Nature News* (June 7, 2017). https://www.nature.com/news/oldest-homo-sapiens-fossil-claim-rewrites-our-species-history-1.22114.

Falk, Darrel R. *Coming to Peace with Science: Bridging the Worlds Between Faith and Biology*. Downers Grove, IL: IVP Academic, 2004.

———. "Theological Challenges Faced by Darwin." In *Darwin, Creation and the Fall: Theological Challenges*, edited by R. J. Berry and T. A. Noble, 75–85. London: Apollos, 2009.

Fergusson, David. *Creation*. Grand Rapids: Eerdmans, 2014.

———. *The Providence of God: A Polyphonic Approach*. Cambridge: Cambridge University Press, 2018.

Finlay, Graeme. *Human Evolution: Genes, Genealogies, and Phylogenies*. Cambridge: Cambridge University Press, 2013.

Florio, Marta, et al. "A Single Splice Site Mutation in Human-specific ARHGAP11B Causes Basal Progenitor Amplification." *Science Advances* 2.12 (2016) E1601941.

Fretheim, Terence E. *Creation Untamed: The Bible, God, and Natural Disasters*. Grand Rapids: Baker Academic, 2010.

———. *God and the World in the Old Testament: A Relational Theology of Creation*. Nashville: Abingdon, 2005.

Fuentes, Agustin. *The Creative Spark: How Imagination Made Humans Exceptional*. New York: Penguin Random House, 2017.

———. *On Being*, November 25, 2020. https://onbeing.org/programs/agustin-fuentes-this-species-moment/

———. "Setting the Stage: Developing the Human Niche Across the Pleistocene." In *Theology and Evolutionary Anthropology: Dialogues in Wisdom, Humility and Grace*, edited by Celia Deane-Drummon and Augustin Fuentes, 13–28. Oxfordshire, UK: Routledge, 2020.

———. *Why We Believe: Evolution and the Human Way of Being*. Foundational Questions in Science. New Haven: Yale University Press, 2019.

Gallup. "40 percent of Americans Believe in Creationism." 2019. https://news.gallup.com/poll/261680/americans-believe-creationism.aspx.

Gee, Henry. *The Accidental Species: Misunderstandings of Human Evolution*. Chicago: University of Chicago Press, 2013.

Gitschlag, Bryan L., et al. "Nutrient Status Shapes Selfish Mitochondrial Genome Dynamics Across Different Leviels of Selection." *Elife* 9 (2020) e56686.

Godinho, R. M., et al. "Supraorbital Morphology and Social Dynamics in Human Evolution." *Nature Ecology and Evolution* 2 (2018) 956–61.

Gould, Stephen Jay. *Rocks of Ages: Science and Religion in the Fullness of Life*. New York: Ballentine, 1999.

Grandview Research. "Biotechnology Market Size: 2022–2030." https://www.grandviewresearch.com/industry-analysis/biotechnology-market.

Green, Joel. "A Wesleyan View." In *Original Sin and the Fall: Five Views*, edited by James B. Stump and Chad Meister, 54–76. Downers Grove, IL: IVP Academic, 2020.

Gruhn, Ruth. "Evidence Grows that Peopling of the Americas Began More than 20,000 Years Ago." *Nature* 584 (2020) 47–48.

Grün, Rainer, et al. "Dating the Skull from Broken Hill, Zambia, and Its Position in Human Evolution." *Nature* 580 (2020) 372–75.

Haarsma, Loren. *When Did Sin Begin? Human Evolution and the Doctrine of Original Sin*. Grand Rapids: Baker Academic, 2021.

Harari, Yuval. *Sapiens: A Brief History of Humankind*. New York: Harper Perennial, 2015.

Hardy, Alister. *The Living Stream: Evolution and Man*. New York: Harper and Row, 1965.

Hare, Brian, and Vanessa Woods. *Survival of the Friendliest: Understanding our Origins and Rediscovering Our Common Humanity*. New York: Random House, 2020.

Harmand, Sonia, et al. "3.3-million-year-old Stone Tools from Lomekwi 3, West Turkana, Kenya." *Nature* 521 (2015) 310–15.

Henrich, Joseph. *The Secret of Our Success: How Culture is Driving Human Evolution, Domesticating Our Species and Making Us Smarter*. Princeton: Princeton University Press, 2015.

Herculano-Houzel, Suzana. *The Human Advantage: A New Understanding of How Our Brain Became Remarkable*. Cambridge: MIT Press, 2016.

Herzog, Werner, dir. *Cave of Forgotten Dreams*. New York: IFC Films, 2010.

Hoffmann, D. L., et al. "U-Th Dating of Carbonate Crusts Reveals Neanderthal Origin of Iberian Cave Art." *Science* 359 (2018) 512–15.

Holliday, Robin, and G. W. Grigg. "DNA Methylation and Mutation." *Mutation Research* 285 (1993) 61–67.

Holt, Jim. "Beyond Belief." *New York Times*, October 22, 2006. https://www.nytimes.com/2006/10/22/books/review/Holt.t.html.

Hrdy, Sarah Blaffer. *Mothers and Others: The Evolutionary Origin of Mutual Understanding*. Cambridge: The Belknap Press of Harvard University Press, 2011.

Hublin, Jean-Jacques, et al. "New fossils from Jebel Irhoud, Morocco and the pan-African origin of *Homo sapiens*." *Nature* 546 (2017) 289–92.

Hurtado, Larry W. *Lord Jesus Christ: Devotion to Jesus in Early Christianity*. Grand Rapids: Eerdmans, 2003.

Hutchinson, Ian. *Monopolizing Knowledge: A Scientist Refutes Religion Denying, Reason Destroying Scientism*. New York: Fias, 2011.

Journal for the Study of the Historical Jesus. 6.2 (2008).

James, William. *The Varieties of Religious Experience: A Study in Human Nature*. New York: Longmans, Green and Co., 1902.

Jobling, Mark, et al. *Human Evolutionary Genetics*. New York: Garland Science, 2014.

John Paul II. *Man and Woman He Created Them: A Theology of the Body*. Boston: Pauline, 2006.

Judson, Horace Freeland. *The Eighth Day of Creation: Makers of the Revolution in Biology*. Commemorative 25th Anniversary ed. Cold Spring Harbor, NY: Cold Spring Harbor Laboratory, 1996.

Kacki, Sasha, et al. "Complex mortuary dynamics in the Upper Paleolithic of the decorated Grotte de Cussac, France." *Proceedings of the National Academy of Sciences* 117 (2020) 14851–56.

Kauffman, Stuart. "Self-organization, Selective Adaptation, and its Limits." In *Evolution at a Crossroads* edited by D. J. Depew and B. H. Web, 169–207. Cambridge: MIT Press, 1985.

Keller, Timothy. *The Reason for God: Belief in an Age of Skepticism*. London: Penguin, 2009.

Keynes, Randall. *Annie's Box*. London: Fourth Estate, 2001.

Kimbel, William H., et al. "Systematic assessment of a maxilla of *Homo* from Hadar, Ethiopia." *American Journal of Physical Anthropology* 103 (1997) 235–62.

Kirschner, Marc W., and John C. Gerhart. *The Plausibility of Life: Resolving Darwin's Dilemma*. New Haven: Yale University Press, 2005.

Kong, Augustine, et al. "Rate of De Novo Mutations and the Importance of Father's Age to Disease Risk." *Nature* 488 (2012) 471–75.

Laland, Kevin. *Darwin's Unfinished Symphony: How Culture Made the Human Mind*. Princeton: Princeton University Press, 2017.

Lane, Nick. *The Vital Question: Energy, Evolution, and the Origins of Complex Life*. New York: W. W. Norton, 2015.

Larson, E. J., and L. Witham. "Leading scientists still reject God." *Nature* 394 (1998) 313–14.

Levenson, Jon D. *Creation and the Persistence of Evil*. Princeton: Princeton University Press, 2013.

Lewis, P. A., et al. "Higher Order Intentionality Tasks Are Cognitively More Demanding." *Social Cognitive and Affective Neuroscience* 12 (2017) 1063–71.

Lewontin, Richard. "Billions and Billions of Demons." *New York Review of Books*. January 9, 1997, 28–32.

Li, Hong Hua, et al. "Induced Chromosome Deletions Cause Hypersociability and other Features of Williams-Beuren Syndrome in Mice." *EMBO Molecular Medicine* 1 (2009) 50–65.

Lodahl, Michael. "The Cosmological Basis for John Wesley's Gradualism." *Wesleyan Theological Journal* 32 (1997) 17–32.

———. *Matthew Matters: The Yoke of Wisdom and the Church of Tomorrow*. Eugene, OR: Cascade, 2021.

Longman, Tremper, and John H. Walton. *The Lost World of the Flood: Mythology, Theology, and the Deluge Debate*. Downers Grove, IL: IVP Academic, 2018.

Losos, Jonathan, B. *Improbable Destinies: Fate, Chance, and the Future of Evolution*. New York: Riverhead, 2017.

Lovejoy, C. Owen, et al. "Combining prehension and propulsion: the foot of *Ardipithecus ramidus*." *Science* 326 (2009) 72–72e8.

Mann, Mark, H. " The Church Fathers and the Two Books Theology." BioLogos, 2012. *https://biologos.org/articles/the-church-fathers-and-two-books-theology*.

Maynard Smith, J., and E. Szathmary. *The Origins of Life: From the Birth of Life to the Origin of Language*. Oxford: Oxford University Press, 1998.

Mayr, Ernest. "Darwin's Influence on Modern Thought." *Scientific American* 283.1 (2000) 79–83.

———. *One Long Argument: Charles Darwin and the Genesis of Modern Evolutionary Thought*. Cambridge: Harvard University Press, 1991.

McCabe, Herbert. *God Matters*. London: Continuum, 2005.

McConville, J. Gordon. *Being Human in God's World: An Old Testament Theology of Humanity*. Grand Rapids: Baker Academic, 2016.

McGrath, Alister E. *Darwinism and the Divine: Evolutionary Thought and Natural Theology*. Hoboken, NJ: Wiley-Blackwell, 2011.

McKnight, Scot, and Dennis R. Venema. *Adam and the Genome: Reading Scripture after Genetic Science*. Grand Rapids: Brazos, 2017.

McLaughlin, Rebecca. *Confronting Christianity: 12 Hard Questions for the World's Largest Religion*. Wheaton, IL: Crossway, 2019.

Middleton, J. Richard. "The Genesis Creation Accounts." In *T & T Clark Handbook of Christian Theology and the Modern Sciences*, edited by John P. Slattery, 15–32. London: T & T Clark, 2022.

———. "Human Distinctiveness and the Origin of Evil." Canadian Scientific and Christian Affiliation Annual Meeting, Langley, Canada: https://www.youtube.com/watch?v=Mop8PQmK-9U, 2018.

———. *The Liberating Image: The Imago Dei in Genesis 1*. Grand Rapids: Brazos, 2005 .

———. "Primal Harmony to a Broken World." In *Earnest: Interdisciplinary Work Inspired by the Life and Teachings of B. T. Roberts,* edited by Andrew C. Koehl and David Basinger, 143–71. Portland, OR: Pickwick, 2017.

Miller, Kenneth. *The Human Instinct: How We Evolved to Have Reason, Consciousness, and Free Will*. New York: Simon and Schuster, 2018.

Moberly, R. W. L. *Old Testament Theology: The Theology of the Book of Genesis*. Ebook ed. Cambridge: Cambridge University Press, 2009.

Moltmann, Jürgen. *God in Creation: A New Theology of Creation and the Spirit of God*. Minneapolis: Fortress, 1993.

———. *The Spirit of Life: A Universal Affirmation*. Translated by Margaret Kohl. Minneapolis: Fortress, 1992.

Monod, Jacques. *Chance and Necessity: An Essay on the Natural Philosophy of Modern Biology*. New York: Collins, 1994.

Moore, John A. *Science as a Way of Knowing: The Foundations of Modern Biology*. Cambridge: Harvard University Press, 1993.

Morrow, Lance. *Evil: An Investigation*. New York: Basic, 2004.

Nachman, M. W., and S. L. Crowell. "Estimate of Mutation Rate per Nucleotide in Humans." *Genetics* 156 (2000) 297–304.

Namba, Takashi, et al. "Human-specific ARHGAP11B Acts in Mitochondria to Expand Neocortical Progenitors by Glutaminolysis." *Neuron* 105 (2020) 867–81.

National Academies of Science and Medicine. "Human Genome Editing: Science, Ethics, and Government." 2017. https://nap.nationalacademies.org/resource/24623/Human-Genome-Editing-highlights.pdf.

National Public Radio. "44,000-Year-Old Indonesian Cave Painting Is Rewriting the History Of Art." *All Things Considered,* December 11, 2019. https://www.npr.org/2019/12/11/786760790/44-000-year-old-indonesian-cave-painting-is-rewriting-the-history-of-art.

Noble, Denis. *Dance to the Tune of Life: Biological Relativity*. Cambridge: Cambridge University Press, 2017.

Nowak, Martin A., and Roger Highfield. *Supercooperators: Altruism, Evolution, and Why We Need Each Other to Succeed*. Edinburgh: Canongate, 2011.

O'Connor, Kathleen. "Why Read the Story in the Garden." In *Pastoral Essays in Honor of Lawrence Boadt, CSP: Reading the Old Testament*, edited by Corrine I. Carvalho, locations 463–720. Ebook ed. Mahwah, NJ: Paulist, 2017.

Oliver, Simon. *Creation: A Guide for the Perplexed*. London: T & T Clark, 2017.

Org., Phys. "First Experimental Genetic Evidence of the Human Self-domestication Hypothesis." *Phys Org,* December 4, 2019, https://phys.org/news/2019-12-experimental-genetic-evidence-human-self-domestication.html.

Orr, H. Allen. "The God Delusion, a Response to D. Dennett's letter." *New York Review of Books,* March 1, 2007, 49.

———. "A Mission to Convert: A Review of God Delusion by Richard Dawkins." *New York Review of Books,* January 1, 2007, 21–24.

Osborn, Ronald E. *Death Before the Fall: Biblical Literalism and the Problem of Animal Suffering*. Downers Grove, IL: InterVarsity, 2014.

Parker, David. "Original Sin: A Fresh Approach." *Evangelical Theology* 13.3 (1989) 51–69.

Pascal, Blaise. *Pensees.* Edited by L. Lafuma. London: Penguin, 1972,

Pearce, Eiluned, et al. "New insights into differences in brain organization between Neanderthals and anatomically modern humans." *Proceedings of the Royal Society B* 280.1758 (2013) 20130168.

Pew Research Center. "Smartphone Ownership Is Growing Rapidly Around the World, but Not Always Equally." https://www.pewresearch.org/global/2019/02/05/smartphone-ownership-is-growing-rapidly-around-the-world-but-not-always-equally/.

Phys. Org. "First experimental genetic evidence of the human self-domestication hypothesis." December 4, 2019.

Pinker, Stephen. *Enlightenment Now: The Case for Reason, Science, Humanism, and Progress.* London: Penguin, 2018.

———. "The Untenability of Faitheism." *Current Biology* 25 (2015) R638–R640.

Plantinga, Alvin. *Where the Conflict Really Lies: Science, Religion, and Naturalism.* New York: Oxford University Press, 2011.

Polkinghorne, John. "Kenotic Creation and Divine Action." In *The Work of Love: Creation as Kenosis*, edited by John Polkinghorne, 90–106. Grand Rapids,: Eerdmans, 2001.

Price, M. "Early humans domesticated themselves, new genetic evidence suggests." *Science* (December 4, 2019) doi:10.1126/science.aba4534.

Provan, Iain. *Seriously Dangerous Religion: What the Old Testament Really Says and Why It Matters.* Waco, TX: Baylor University Press, 2014.

Purugganan, Michael D. "What is Domestication?" *Trends in Ecology and Evolution* 37.8 (2022) 663–71.

Quammen, David. *The Reluctant Mr. Darwin: An Intimate Portrait Of Charles Darwin and the Making of His Theory of Evolution.* New York: Norton, 2006.

Quiles, A., et al. "A High-precision Chronological Model for the Decorated Upper Paleolithic Cave of Chauvet-Pont d'Arc, Ardèche, France." *Proceedings of the National Academy of Sciences* 113 (2016) 4670–75.

Ramm, Bernard. *Offense to Reason: The Theology of Sin.* New York: Harper and Row, 1985.

Richter, Clemens. "Remembering Johann Gregor Mendel: a human, a Catholic priest, an Augustinian monk, and abbot." *Molecular Genetics and Genomic Medicine* 3 (2015) 384–86.

Ritchie, Sara Lane. *Divine Action and the Human Mind.* Cambridge: Cambridge University Press, 2019.

Rutledge, Fleming. *The Crucifixion: Understanding the Death of Jesus Christ.* Grand Rapids: Eerdmans, 2017.

Safina, Carl. *Becoming Wild: How Animal Cultures Raise Families, Create Beauty, and Achieve Peace.* New York: Picador, 2020.

———. *Beyond Words: What Animals Think and Feel.* New York: Henry Holt, 2015.

Scerri, E. M. L., et al. "Did Our Species Evolve in Subdivided Populations Across Africa, and Why Does It Matter?" *Trends in Ecology and Evolution* 33 (2018) 582–94.

Schmidt, E. R. E., et al. "A Human-specific Modifier of Cortical Connectivity and Circuit Function." *Nature* 599 (2021) 640–44.

Shermer, Michael. *Why Darwin Matters: The Case Against Intelligent Design*. Austin, TX: Holt McDougal, 2007.

Shilton, Dor, et al. "Human Social Evolution: Self-Domestication or Self-Control?" *Frontiers in Psychology* (2020) 134.

Smith, James K. A. "Is the Universe Open for Surprise? Pentecostal Ontology and the Spirit of Naturalism." *Zygon* 43 (2008) 879–96.

Snyder, Howard A. "Salvation Means Creation Healed." Unpublished paper presented at "the Kingdom Conference," Baltimore, 2005.

Solomon, Sheldon, et al. *The Worm at the Core: On the Role of Death in Life*. New York: Random House, 2015.

Stirrat, Michael, and R. Elizabeth Cornwell. "Eminent scientists reject the supernatural: a survey of the Fellows of the Royal Society." *Evolution: Education and Outreach* 6 (2013) 33.

Stringer, Chris, and Julia Galway-Witham. "On the Origin of Our Species." *Nature* 546 (2017) 212–14.

Stump, J. B., and Chad Meister. *Original Sin and the Fall: Five Views*. Downers Grove, IL: IVP Academic, 2017.

Seuss, Dr. *Oh, The Places You'll Go*. New York: Random House, 1990.

Swamidass, Joshua. *The Genealogical Adam and Eve: The Surprising Science of Universal Ancestry*. Downers Grove, IL: InterVarsity, 2019.

Sykes, Rebecca Wragg. *Kindred: Neanderthal Life, Love, Death and Art*. London: Bloomsbury Sigma, 2020.

Tanner, Kathryn. "Creation and Providence." *The Cambridge Companion to Karl Barth*, edited by John Webster, 111–26. Cambridge: Cambridge University Press, 2000.

Tassabehji, May et al. "GTF2IRD1 in Craniofacial Development of Humans and Mice." *Science* 310 (2005) 1184–87.

Tattersall, Ian. *Masters of the Planet: The Search for Our Human Origins*. New York: St. Martin's, 2012.

———. *The Strange Case of Rickey Cossack and Other Cautionary Tales from Human Evolution*. London: Palgrave Macmillan., 2015.

Tattersall, Ian, and Rob DeSalle. *The Accidental* Homo sapiens*: Genetics, Behavior and Free Will*. New York: Pegasus, 2019.

Taylor, John V. *The Go-Between God*. Minneapolis: Fortress, 1972.

Taylor, Thomas. *The Rhetoric, Poetics, and Nicomachean Ethics of Aristotle, Vol. II*. London: James Black and Son, 1818.

Tomasello, Michael. *Becoming Human: A Theory of Ontogeny*. Cambridge: The Belknap Press of Harvard University Press, 2019.

———. *A Natural History of Human Morality*. Cambridge: Harvard University Press, 2016.

Tomasello, Michael, et al. "Understanding and Sharing Intentions: The Origins of Cultural Cognition." *Behavioral and Brain Sciences* 28 (2005) 675–735.

Trinkaus, Erik, and Alexandra P. Buzhilova. "Diversity and differential disposal of the dead at Sunghir." *Antiquity* 92.361 (2018) 7–21.

U.S. Dept. of Commerce: Office of Technology Policy. "Meeting the Challenge (U.S. Industry Faces The 21st Century): The U.S. Biotechnology Industry." July 1997. https://webharvest.gov/peth04/20041017062206/http://www.technology.gov/Reports/biotechnology/cd93a.pdf.

Van Wolde, Ellen. *A Semiotic Analysis of Genesis 2–3: A Semiotic Theory and Method of Analysis Applied to the Story of the Garden of Eden*. Assen, the Netherlands: Van Gorcum, 1989.

Varki, Ajit. "Did Human Reality Denial Breach the Evolutionary Psychological Barrier of Mortality Salience? A Theory that Can Explain Unusual Features of the Origin and Fate of Our Species." In *Evolutionary Perspectives on Death*, edited by T. Shackelford and V. Zeigler-Hill, 109–35. Cham, Switzerland: Springer, 2019.

———. "Why Are There No Persisting Hybrids of Humans with Denisovans, Neanderthals, or Anyone Else?" *Proceedings of the National Academy of Sciences. U.S.A* 113.17 (2016) E2354.

Varki, Ajit, and Danny Brower. *Denial: Self Deception, False Beliefs, and the Origins of the Human Mind*. New York: Twelve, 2014.

Von Balthasar, Hans U. *Prayer*. Ebook ed. San Francisco: Ignatius, 1986.

Wagner, Andreas. *Arrival of the Fittest: Solving Evolution's Greatest Puzzle*. Carmel, IN: Current, 2014.

Wall Kimmerer, Robin. *Gathering Moss: A Natural and Cultural History of Mosses*. Corvallis, OR: Oregon State University Press, 2003.

Walton, John H. *The Lost World of Adam and Eve: Genesis 2–3 and the Human Origins Debate*. Downers Grove, IL: InterVarsity, 2015.

———. *The Lost World of Genesis One: Ancient Cosmology and the Origins Debate*. Downers Grove, IL: InterVarsity, 2010.

Walton, John H., and Brent Sandy. *The Lost World of Scripture: Ancient Literary Culture and Biblical Authority*. Downers Grove, IL: InterVarsity, 2013.

Walton, John H., and Tremper Longman III. *How to Read Job*. Downers Grove, IL: IVP Academic, 2015.

Ward, Keith. *Why There Almost Certainly Is a God: Doubting Dawkins*. Kent, UK: Lion House, 2008.

Webster, John. "Providence." In *Church Dogmatics: Reformed Theology for the Church Catholic*, edited by Michael Allen, and Scott Swain, 148–65. Grand Rapids: Baker Academic, 2016.

———. "Trinity and Creation." *International Journal of Systematic Theology* 12 (2010) 4–19.

White, Randall, et al. "Still no archaeological evidence that Neanderthals created Iberian cave art." *Journal of Human Evolution* 144 (2019) 10264.

Wilkins, Adam A., et al. "The 'Domestication Syndrome' in Mammals: A Unified Explanation Based on Neural Crest Cell Behavior and Genetics." *Genetics* 197 (2014) 795–808.

Williams, Rowan. *Christ: The Heart of Creation*. London: Bloomsbury Continuum, 2018.

Wilson, David Sloan. *This View of Life: Completing the Darwinian Revolution*. New York: Pantheon, 2019.

Wilson, David Sloan, and Edward O. Wilson. "Rethinking the Theoretical Foundation of Sociobiology." *Quarterly Review of Biology* 82 (2007) 327–48.

Wilson, Edward O. *The Meaning of Human Existence*. New York: Liveright, 2014.

———. *The Social Conquest of the Earth*. New York: Liveright, 2012.

Wohlleben, Peter. *The Hidden Life of Trees: What They Feel, How They Communicate*. Vancouver: Greystone, 2016.

Wrangham, Richard. *Catching Fire: How Cooking Made Us Human*. New York: Basic, 2009.

———. *The Goodness Paradox: The Strange Relationship between Peace and Violence in Human Evolution*. New York: Vintage, 2019.

———. "Hypotheses for the Evolution of Reduced Reactive Aggression in the Context of Human Self-Domestication." *Frontiers in Psychology* 10 (2019) 1914.

Wright, N. T. *The Day the Revolution Began: Reconsidering the Meaning of Jesus's Crucifixion*. San Francisco: HarperOne, 2016.

———. *History and Eschatology: Jesus and the Promise of Natural Theology*. Waco, TX: Baylor University Press, 2019.

———. *The Resurrection of the Son of God*. Minneapolis: Fortress, 2003.

———. *Surprised by Scripture: Engaging with Contemporary Issues*. London: SPCK, 2014.

Yong, Amos. "Nature Laws and Divine Intervention: What Difference Does Being Pentecostal or Charismatic Make?" *Zygon* 43 (2008) 961–89.

Yue, F., et al. " A Comparative Encyclopedia of DNA Elements in the Mouse Genome." *Nature* 515 (2014) 355–64.

Zanella, Matteo, et al. "Dosage Analysis of the 7q11.23 Williams Region Identifies BAZ1B as a Major Human Gene Patterning the Modern Human Face and Underlying Self-Domestication." *Science Advances* 5 (2019) eaaw7908.

Zimmer, Carl. "CRISPR, 10 Years On: Learning to Rewrite the Code of Life." *New York Times*, June 27, 2022.

———. "What Makes Your Brain Different From a Neanderthal's?" *New York Times*, September 8, 2022.

www.ingramcontent.com/pod-product-compliance
Lightning Source LLC
LaVergne TN
LVHW102012210926
848483LV00011B/324
* 9 7 8 1 6 6 6 7 5 7 0 2 6 *